SOCIAL FACTORS
IN MENTAL HEALTH
AND ILLNESS

Research in Community and Mental Health

Joseph P. Morrissey, Series Editor

CONTENTS

PART III. METHODS

LIST OF CONTRIBUTORS

Mitch Berbrier   University of Alabama in Huntsville

David E. Biegel   Case Western Reserve University

Barbara J. Burns   Duke University Medical Center

William W. Eaton   Johns Hopkins University

Daniel E. Ford   Johns Hopkins University School
                                    of Medicine

Donald K. Freeborn   Center for Health Research,
                                    Portland, Oregon

Linda K. George   Duke University

Stephen M. Goldfinger   State University of New York,
                                    Health Science Center, Brooklyn

Marilyn J. Kennedy   Case Western Reserve University

Anthony C. Kouzis   Johns Hopkins University School
                                    of Medicine

Joseph P. Morrissey   University of North Carolina at
                                    Chapel Hill

John P. Mullooly   Center for Health Research
                                    Portland, Oregon

Elizabeth A. R. Robinson   Alcohol Research Center
                                    Ann Arbor, Michigan

Mark Schlesinger   Yale University School of Medicine

Aileen Schulte   State University of New York
                                    at New Paltz

Russell K. Schutt   University of Massachusetts
                                    at Boston

Diana Shye   Ben-Gurion University of the
                                       Negev, Israel

William H. Sledge   Yale University School of Medicine

Jeffery W. Swanson   Duke University Medical Center

Marvin Swartz   Duke University Medical Center

Neil M. Thakur   University of North Carolina at
                                    Chapel Hill

Maxine Seaborn Thompson   North Carolina State University at
                                    Raleigh

PART I

SOCIAL INTEGRATION AND HELP-SEEKING

BINDING AND NONBINDING INTEGRATION:

THE RELATIONAL COSTS AND REWARDS OF SOCIAL TIES ON MENTAL HEALTH

Mitch Berbrier and Aileen Schulte

ABSTRACT

This paper explores the relationship between social integration and mental health. We extend the notions of relational costs and relational rewards (Levinger and Huesmann 1980), and we introduce the concepts of binding and nonbinding social integration. Binding social ties include obligatory relations, such as those with family or at work; nonbinding integration represents more voluntary relationships, such as those with friends, neighbors, or membership in various organizations. We argue (1) that the direct effects of social integration on psychological distress arise out of the more proximate effects of relational costs and rewards, and (2) that costs are more associated with binding integration, while rewards are more associated with nonbinding integration. We test these hypotheses using a two wave panel study of individuals from the Indianapolis area ($N = 486$) and find some support for our claims. Most clearly we find that the binding nature of relationships seems to effect mental health outcomes. Because of inherent limitations

Research in Community and Mental Health, Volume 11, pages 3-27.
Copyright © 2000 by JAI Press Inc.
All rights of reproduction in any form reserved.
ISBN: 0-7623-0671-8

regarding how costs and rewards of relationships have been measured—both tra-
ditionally and in our data—support for our ideas about relational costs and
rewards is more tenuous and will require further investigation. We conclude that
the main effect of social integration may fruitfully be addressed by considering
the binding or nonbinding nature of social ties as the source of relational costs and
rewards which directly affect mental health.

INTRODUCTION

One of the major issues in stress research is the relationship between social support and mental health. However, as several researchers have noted (e.g., Cohen
and Wills 1985; House, Umberson, and Landis 1988; Mirowsky and Ross 1986),
unclear definitions of social support have often equated social integration (the
number and frequency of social contacts) with the receipt or perception of social
support. Mapping these distinctions has resulted in a great deal of conceptual clar-
ification, but also a growing list of terms, some with very similar meaning. For
example, Cohen and Wills (1985) distinguish between structural and functional
forms of support; House and colleagues (1988) between embeddedness and support; and Mirowsky and Ross (1986) between social integration and social support. In this paper we use the term "social integration" to refer to the number and
frequency of social contacts, and "social support" to refer to the perception or
receipt of support from those contacts. Although the recent trend in empirical
studies has been to differentiate between the existence of social ties and social
support (see, e.g., the discussions in Dean, Kolody and Wood 1990, p. 149;
Loscocco and Spitze 1990, p. 315; Turner, Grindstaff, and Phillips 1990, p. 47),
the resulting focus on perceived social support has glossed over the mental health
implications of social integration. Furthermore, the existence of social ties does
not always entail social support (Bullers 1997; Steinberg 1995; Voight 1996).
Indeed, chronic strains, negative life events and other stressors are a possible
result of any social tie (Rook 1984, 1992; Pagel, Erdly, and Becker 1987). Given
that the existence of social ties is both temporally prior to and analytically distinct
from eliciting support from those ties, the relationship between social integration
and mental health merits separate consideration from the investigation of
perceived social support.

In this paper we conceptualize social integration as similar to Cohen and Wills'
(1985) idea of "global" integration, referring to the totality of the range of social
ties, rather than to particular relationships (e.g., Hughes and Gove 1989; Umberson 1987; Umberson and Gove 1989). Similarly, we draw from the work of network theorists who tend to employ global measures. For example, Haines and
Hurlbert (1992) distinguish between studies of the "dyadic approaches" which
study the traits of contacts and ties (Wellman and Wortley 1990) and their own
network structure approach assesses the impact of the traits of social networks,

such as size, density, and range. Our study employs a simple measure of social networks—size and frequency of contact—and employs a dyadic approach, theorizing on the characteristics of social ties. Social integration is a precursor to several determinants of mental health—some of which may be positive and some negative. Our goal in this article is to disaggregate the costs and benefits of social ties—which we call *relational costs* and *relational rewards* (Levinger and Huesmann 1980)—in order to explain the main effect of integration on mental health. To disentangle these effects, we argue that relationships with close ties and many responsibilities, which we call *binding integration*, should be examined in a different light from those with more distant ties and fewer responsibilities, which we term *nonbinding integration*. We propose that, relatively speaking, binding integration is more closely associated with relational costs and nonbinding integration is more closely associated with relational rewards. Thus, binding integration is more likely to result in negative consequences for mental health, while nonbinding integration is more likely to have positive effects. The well-documented main effect of social integration on mental health (Bell, LeRoy, and Stephenson 1982; Lin, Ensel, Simeone, and Kuo 1979; Miller and Ingham 1979; Williams, Ware, and Donald 1981) can then be understood as the cumulative effect of the relational costs and rewards of binding and nonbinding social relationships.

THE EFFECTS OF SOCIAL INTEGRATION

Relational Rewards: The Positive Effects of Social Integration

According to Levinger and Huesmann (1980), relational rewards accrue automatically as a product of social integration, as opposed to behavioral rewards, which are contingent upon specific behavioral exchanges between individuals. Thoits (1985, p. 57ff) suggests three ways that relational rewards may be beneficial: as a set of identities or a "sense of belonging," as a source of positive self-evaluation or reflected self-esteem, and as a basis for a sense of mastery. Similarly, several studies have linked either self-esteem or mastery, or both, to psychological health outcomes in the context of both negative life events and chronic strains (e.g., Kessler and Essex 1982; Pearlin, Menaghan, Lieberman, and Mullan 1981; Pearlin and Schooler 1978; Turner and Noh 1983).

Another frequently suggested product of social integration is perceived social support. Several studies have found relationships between measures of social integration and social support (e.g., Cutrona 1986; Sarason, Shearin, Pierce and Sarason 1987; Schaefer, Coyne and Lazarus 1981), and the assumption has sometimes been that social support is one of the possible outcomes of the existence of social ties (House et al. 1988; Wellman and Wortley 1990). While some of these studies have argued that self-esteem and perceived social support have a buffering effect—that is, that they each have psychological benefits which are called into effect in response to either a negative life event or a particular strain (Cohen and Wills 1985)—our approach is to assess the degree to which these variables are associated with either binding or nonbinding integration (discussed below).

Our hypothesis is that *the main effect of social integration on psychological health* (Bell, LeRoy, and Stephenson 1982; Lin, Ensel, Simeone, and Kuo 1979; Miller and Ingham 1979; Williams, Ware, and Donald 1981) *can be explained, in part, by the relation of social integration to those psychological variables that have stronger and more direct positive effects on mental health.* We suggest that some of the relational rewards of social integration are social support, self-esteem, mastery, and senses of identity and belongingness.

Relational Costs: The Negative Effects of Social Integration

Distinguishing conceptually between social integration and social support enables us to avoid the questionable assumption that having a relationship is equivalent to getting support (Barrera 1986; Pescosolido and Georgianna 1989; Schaefer et al. 1981). Indeed, there exists a burgeoning literature on the negative effects of social ties (e.g. Bullers 1997; Coyne and DeLongis 1986; Rook 1984, 1992). Specifically, Rook's research (1984, 1992) challenges the notion that social ties provide social support; she has found that negative social interactions are more consistently and strongly related to psychological states than positive ones. Similarly, Fiore, Becker, and Coppel (1983) found that "the extent of upset with the social network" was a much better predictor of depression than the extent of helpfulness of the network (p. 433), and Milardo (1987), who reviews the literature on the effects of social networks following divorce, posits "interference"— disapproval, criticism, sabotage, or the withholding of support—as a cost that is positively correlated with the size of the social network.

These findings indicate that while there is a direct main effect of integration on mental health, it does not follow that all ties are always experienced as positive: aspects of social integration can also have negative effects. We propose that some of the mechanisms through which social integration may negatively affect well-being are stressors, namely chronic strains and negative life events which, using Levinger and Huesmann's (1980) term, we call *relational costs.* The relationship of these relational costs to psychological distress has been well documented (for chronic strains, see Ensel and Lin 1991; Kessler and McLeod 1985; Pearlin and Lieberman 1979; for negative life events, see Thoits 1983). Thus in contrast to the "positive" side of the argument, we hold that *social integration also increases the risk of experiencing chronic strains and negative life events which in turn increases psychological distress.*

Figure 1 presents a model of the social integration and mental health that incorporates these claims: *the main effect of social integration on psychological well-being is understood as the cumulative effects of relational costs and rewards.* Relational rewards represent the positive, or stress-buffering, aspects of social

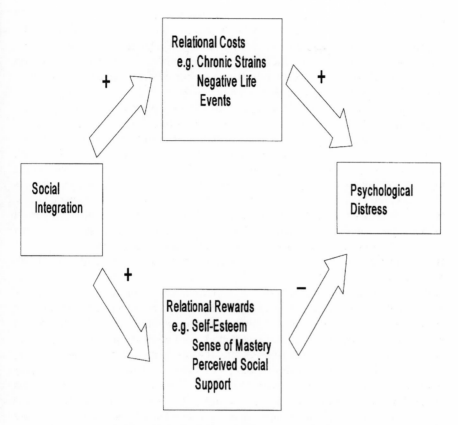

Figure 1. The Main Effect of Social Integration Explained by
the Mediation of Relational Costs and Rewards

integration that decrease psychological distress. Relational costs represent the
negative, or stress-exacerbating, influences of social integration that increase psychological distress.

RETHINKING SOCIAL TIES: BINDING AND
NONBINDING INTEGRATION

While the model in Figure 1 accounts for both the rewards and costs of social relationships, it does not consider the different advantages and obligations of the
diverse types of social ties. That is, the concept of integration presumes a number
of heterogenous ties which may have independent effects. While it would not be

parsimonious to model the effect of each independent relationship, a number of ties may have similar effects due to the nature of their obligations. Specifically, certain kinds of ties may be more closely related to relational costs while others are related to relational rewards. As a result, these different kinds of ties will also have different effects on psychological well-being. We call these different forms of social ties *binding and nonbinding integration.*

Binding integration refers to relationships to which people feel bound for emotional, financial, or other reasons. While people often experience rewards from these relationships, they also often entail obligations and stresses that cannot be avoided. Thus in our view binding ties most likely prevail in the family and work domains. Chronic strains arising in the work or family roles have long been noted to have negative effects upon well-being and to increase psychological distress (for work, see House 1981; for family, see McLanahan and Adams 1987 or Ross, Mirowsky, and Huber 1983). Stress within or losses of these relationships most often lead to negative feeling states (Thoits 1985). In fact, when researchers focus on specific sources of stress they most commonly focus on family and work (e.g., Bolger, DeLongis, Kessler, and Wethington 1989; Eckenrode and Gore 1990; Voydanoff and Donnelly 1989).

We argue that family members and work associates represent ties that can be very "tight" in that they cannot easily be broken, even temporarily. People are "stuck" with their families, and in our society it is, perhaps, a moral imperative to maintain ties of loyalty with family members even when people do not get along or even like each other. Furthermore, cultural analysis indicates that this collectivist ethic might also clash with the increasing individualism of people in modern societies (Bellah, Masden, Sullivan, Swidler, and Tipton 1985), exacerbating this effect. For these reasons, among others, children, parents, in-laws, siblings, and other relatives often are potentially significant sources of stress. Similarly, we are bound to relationships at work, be they with bosses, coworkers, subordinates, or clients. In this case, the binding nature of the relationship may not come from cultural or moral constraints, but from financial ones.

Whether the obligations arise for moral, cultural, or financial reasons, binding integration captures the nature of family and work relationships: these social ties inherently involve an extensive commitment of time and energy that is not easily averted. Due to these substantial constraints, binding integration seems likely to be associated with relational costs. Following the logic of Figure 1, it seems unlikely that these "binding" features of social integration will explain the established beneficial effects of integration on psychological states; if anything they would be expected to entail harmful effects.

Nonbinding integration, on the other hand, refers to those relationships in which people engage more voluntarily. In comparison to family and work, relationships with friends, club or church associates tend to be more "loose" in that they are of a continuously more voluntary nature, and can more easily be broken off for shorter or longer periods of time. People engage in these relationships because they want to; there are fewer and weaker external constraints compelling these sorts of relations, and they fit the voluntary nature of the contemporary individualist ethos (Bellah et al. 1985). Given their voluntary nature, it is possible that these relationships involve fewer relational costs, and generate many of the relational rewards that are produced by social integration. In other words, nonbinding integration seems more likely to be associated with reduced distress than binding integration.

These ideas resonate well with some recent literature on social relationships and mental health. In a discussion of women's adjustment to divorce, for example, Milardo (1987) noted that friendships provide companionship, recreation, and intimacy without the "burden," "complexity," or "threats of judgment" often associated with "responsibilities of motherhood, obligations to kin, and constraints of wage work" (p. 90). Similarly, in Morgan's (1989) discussion of adjustments to widowhood, he described how widows could forego relationships that did not meet personal standards of support, and instead pursue relationships that were supportive. He concluded that "turning from family to nonfamily relationships means turning from a pattern of commitment to one of flexibility" (Morgan 1989, p. 106). Findings such as these lend credence to our claim that relationships with differing obligations should be investigated separately in an examination of the mental health consequences of social integration.

Before we turn to the full model, it is important to emphasize that the argument being made here is not that some relationships are devoid of costs and others of rewards. Clearly there are many benefits associated with family and work relationships, as well as costs involved in nonbinding relations. For instance, family and work roles provide a sense of identity and a possibility for positive self-evaluation, much in the way suggested by Thoits (1985). Also, spouses and other family members may provide social support in response to life strains or negative life events (Jackson 1992; Umberson 1987). Undoubtedly, family ties can have both costs and benefits (Umberson and Gove 1989). Similarly, there is evidence to indicate that friend and neighbor relationships are not purely rewarding (Rook 1984). In this paper we are arguing only that to the degree that relationships are obligatory they will be less rewarding and less likely to have a beneficial effect on mental health, and to the degree that they are voluntary they will be more rewarding and more likely to have a beneficial effect on mental health.

MODEL AND HYPOTHESES

The final version of the model is presented in Figure 2. This model adds to the first by proposing that it is the degree of "bindingness" which explains the relationship between social integration and psychological distress, and also that binding relationships will be most strongly associated with relational costs and that

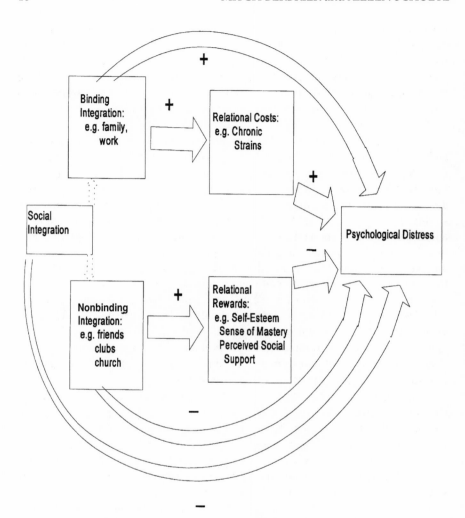

Figure 2.   The Main Effect of Social Integration on Psychological Distress
Explained by Breaking It into Binding and Nonbinding Forms, with
Binding Integration More Closely Associated with Relational Costs and
Nonbinding Integration More Closely Associated with Relational Rewards

nonbinding relationships will be most strongly associated with relational rewards.
The ideas embodied in Figure 2 can be summarized in the following hypotheses:

Hypothesis 1.   Social integration has a negative effect on psychological distress.

Hypothesis 2.   Social integration can be fruitfully broken down into two components: binding integration and nonbinding integration. The nonbinding integration component will have a negative effect on psychological distress—accounting for the relationship in Hypothesis 1—but the binding integration component will not show this effect, and may even increase distress.

Hypothesis 3.   While social integration will have positive effects on both relational rewards and costs, relatively speaking nonbinding integration will be more closely associated with rewards (Hypothesis 3a) and binding integration with costs (Hypothesis 3b).

Hypothesis 4.   Relational rewards have a stress-buffering effect on psychological distress while the effects of relational costs are stress-exacerbating. While this effect is well-documented in the literature (i.e., chronic strains increase psychological distress [e.g., Pearlin and Lieberman 1979], and self-esteem is associated with lower levels of psychological distress [e.g. Cohen and Wills 1985]), here we specifically consider the relationship of these concepts to binding and nonbinding integration.

What follows is an initial empirical test of these claims, using a two-wave panel study which provides us with measures of social integration, binding and nonbinding integration, psychological distress, two of the four suggested relational rewards (perceived social support and self-esteem), and one of the two suggested relational costs (chronic strains). In addition, one other relational reward specifically associated with binding integration roles (perceived role performance) is included, to account for potential benefits from these social ties.

METHODS

The data used in these analyses are from a two-wave panel study (Thoits 1991) designed to investigate the mental health implications of chronic strains and life events in various role domains. A stratified random sample of married and divorced individuals from the Indianapolis area were selected through random digit dialing techniques and systematic sampling procedures of state records of divorces. Married respondents were obtained solely through random digit dialing techniques, while divorced respondents were selected through both means (see Simon 1992 or Thoits 1991 for a more extensive explanation of data collection techniques). Structured personal interviews were conducted in 1988 and 1990 (N = 700 in the first wave). A total of 527 respondents participated in both waves of the study, yielding an attrition rate of 24 percent. The average duration of the interviews was one hour and 40 minutes for the first wave and one hour and 20 minutes for the second. In comparison to a national probability sample of married and divorced individuals, these respondents were demographically similar to those living in other U.S. cities.

For the purposes of these analyses, individuals who were married or employed at time 1 and had lost that tie by the time of the second interview (due to divorce or unemployment) were eliminated from our sample. Data from these respondents would have been meaningless since they would have been asked about their levels of integration but would not have been asked about relational costs or rewards. With the exclusion of these respondents, the final sample size was 486.

Measures

In order to measure the long-term effects of social integration, all measures of binding and nonbinding integration (as well as the control variables) were taken from the first wave of the panel study; all relational rewards and chronic strains (relational cost), as well as measures of psychological distress, were taken from the second wave. While our main goal here was to separate integration measures in time from distress measures, we nevertheless recognize that doing so does not establish a time order of the independent (integration) and dependent (distress) variables. In order to ascertain the effect of these choices on the data, we ran two other analyses: one with control variables measured at time 2; and one with relational costs and rewards measured at time 1. We did not find any significant differences between these analyses and those presented here.

Social Integration

Measures of social integration were created using data on numbers of social ties and frequencies of social contacts. The measure assessed integration with one's children, parents, in-laws, other relatives, associates at work, neighbors, friends, church colleagues, hobby or leisure group members, and other group or organization members. Since the original items differed in range of responses, it was necessary to develop a way to combine items measured on different scales. To this end, a number of different combinations were explored. Specifically, in one analysis we standardized the measures; in another we reduced each of the measures to a three-point scale (to represent low, moderate, and high levels) and we also examined these as dichotomous measures accounting only for the presence or absence of each social bond. These various measures did not yield radically different results. The measures used in this article were chosen because they preserve the complexity of the original items and include more extensive information regarding frequency of contacts.

The measures for both binding and nonbinding integration are presented in the Appendix. When a respondent did not have the social tie in question (i.e., integration with persons at work for those who are unemployed), social integration in that role domain was coded as a zero. Measures of frequency of contacts with family were summed to form an index of binding integration, and measures regarding contacts with friends, neighbors, and religious, leisure or other groups were summed to form an index of nonbinding integration. It is important to note that the items provide information from different role domains that may be completely unrelated. Therefore, the items are not expected to have shared variances and the resulting alpha reliabilities are relatively low—for binding integration, 0.26, for nonbinding integration, 0.30, and 0.26 for all integration.

Relational Costs and Rewards

We have hypothesized that nonbinding integration is associated with a number of psychological benefits or "relational rewards." In these analyses, we include self-esteem and sense of mastery as indicators of relational rewards, expecting these rewards to be associated with non-binding integration, but not with binding integration. However, we will of course examine the relationships between both forms of integration and all relational costs and rewards. Self-esteem, or positive self-attitudes, was measured with Rosenberg's Self-Esteem Scale (1979); mastery, or sense of control over one's life, was measured with Pearlin and Lieberman's Sense-of-Mastery Scale (1979). Level of agreement was measured on a four-point scale from "strongly agree" to "strongly disagree." On each scale, items which indicate negative feelings toward one's self were backcoded so that a high score on the scale indicates positive attitudes. Both scales have demonstrated good construct validity (for self-esteem, see Rosenberg 1979; for mastery, see Pearlin and Schooler 1978). In these analyses, the alpha reliabilities are .85 for the self-esteem scale and .73 for the sense of mastery scale.

A scale which measured perceived social support from a confidante was deemed to be a relational reward in the realm of nonbinding integration. Items for this scale were derived from scales developed by Pearlin and Lieberman (1979) and Procidano and Heller (1983). Responses ranged on a four-point scale from "strongly agree" to "strongly disagree." The alpha reliability for this scale was .94.

The perceived role performance measure was derived from a series of questions which asked "how good" the respondents felt they were in the roles of parent, spouse, worker, son/daughter, and son/daughter-in-law. Responses ranged from: 1 = "extremely poor/unsuccessful" to 7 = "extremely good/successful." If the respondent did not have the corresponding social tie, he/she was given a score of zero for the perceived role performance in that role domain. Items were summed to form an index of the perceived role performance across roles—a relational reward of binding integration. The alpha reliability for this index is .58 (but again, the items provide information from different role domains that may be completely unrelated, and are not expected to have shared variances).

Regarding our measurement of relational costs, the chronic strains measure was derived from a series of questions which asked "how stressful" it was to hold the roles of parent, spouse, worker, son/daughter, and son/daughter-in-law. Responses ranged from "strongly agree" to "strongly disagree" on a four-point scale. Respondents who did not have the social tie in question were given a score of zero for chronic strains in that role domain. Items were summed to form an index of chronic strains, which are relational costs of binding integration. The alpha reliability for the chronic strains index, also derived from different domains, is .40.

For both perceived role performance and the chronic strains indices we assign zeroes for cases in which a respondent does not occupy the role in question, since the absence of a role precludes deriving costs or benefits from it. For example, respondents who claim that their job is "not at all stressful" are assumed to have higher levels of perceived work stress than those respondents who are unemployed or homemakers. Since we have controlled for the presence or absence of roles measured in the chronic strains and perceived role performance indices, this method of coding does not bias our results.

Psychological Distress

The distress measure consisted of 23 items selected from the anxiety, depression and somatization subscales of the Brief System Inventory (BSI) (Derogatis and Spencer 1982). The overall distress score used in the analyses is an aggregation of the items from the three subscales. In this study, these measures generally exhibited high reliability. Cronbach's alpha was relatively high for each of the subscales (for depression .77; for anxiety .78; and for somatization .72), yielding an overall alpha reliability of .87.

Control Variables

All regression analyses control for the following variables: gender (female = 1, male = 0), age, marital status (married = 1; divorced/widowed = 0), education (measured on a five-point ordinal scale), total family income (measured on a 21 point ordinal scale), employment (employed = 1; unemployed/retired/homemaker = 0) and race (black/African American = 1; other = 0). Due to the possible confounds introduced by the summation of items on the binding integration, chronic strains, and perceived role performance indices, we include a control variable ranging from one to five, which sums the number of binding roles held by the respondent. This variable is included in regression analyses of the full theoretical model to control for the cumulative nature of the chronic strains and perceived role performance indices.

Table 1. Correlations of All Variables ($N = 486$)

	1	2	3	4	5	6	7	8	9	10	11	12	13	14	15	16
1. All Integration	1.0															
2. Binding Integration	.71***	1.0														
3. Nonbinding Integration	.72***	.02	1.0													
4. Chronic Strains	.40***	.55***	.02	1.0												
5. Perceived Role Performance	.48***	.62***	.06	.73***	1.0											
6. Self-Esteem	.14**	.03	.16***	-.09*	.18***	1.0										
7. Sense of Mastery	.03	.01	.04	-.10*	.08+	.66***	1.0									
8. Perceived Social Support	.20***	.01	.27***	-.13**	-.05	.21***	.19***	1.0								
9. Gender (F = 1)	.06	.02	.06	-.03	-.11*	-.11*	-.16***	.21***	1.0							
10. Marital Status (Marr. = 1)	.39***	.44***	.12*	.63***	.68***	.09+	-.03	-.12**	-.15**	1.0						
11. Employment (Yes = 1,No = 0)	.09+	.18***	-.05	.15**	.22***	.08+	.10*	-.01	-.14**	-.08+	1.0					
12. Total Family Income	.22***	.15*	.16**	.32***	.40***	.19***	.08+	.00	-.17***	.01	.01	1.0				
13. Education	.15*	-.01	.22***	.08+	.11*	.16***	.10*	.04	-.02	.10*	.10*	.34***	1.0			
14. Race (African Amer. = 1)	-.07	-.11*	.01	-.19***	-.16***	-.08+	.09+	.03	.08+	-.04	-.05	-.19***	-.16***	1.0		
15. Age	-.19***	-.40***	.12*	-.31***	-.30***	-.06	-.06	-.05	-.04	-.19***	-.19***	.08+	.01	.05	1.0	
16. Psychological Distress	-.14**	-.04	-.16***	.09+	-.15*	-.57***	-.43***	-.17**	.20***	-.10*	-.13*	-.10*	-.11**	.01	.17***	1.0

Notes: $^+ p \leq 0.10$
$^* p \leq 0.05$
$^{**} p \leq 0.01$
$^{***} p \leq 0.001$

15

RESULTS

Table 1 presents the correlations among all variables in the study. These basic results were encouraging. Considering all types of social contacts, all integration is significantly and negatively related to psychological distress ($r = -.14, p < .01$). This confirms our first hypothesis, and replicates the common finding in the literature that a greater number of social contacts is associated with lower levels of psychological distress. However, when binding and non-binding integration are considered separately, nonbinding integration is significantly and negatively related to psychological distress ($r = -.16, p < .001$), while the relationship between binding integration and psychological distress is not significant. This is consistent with Hypothesis 2: nonbinding integration, operationalized as greater contact with friends, neighbors, or in organized groups, is associated with lower levels of psychological distress; binding integration—greater contacts with family members or at work—is unrelated to psychological distress.

Binding and nonbinding social ties also exhibit differential relationships with relational costs and rewards. Nonbinding integration is significantly related to both self-esteem ($r = .16, p < .001$) and perceived social support ($r = .27, p < .001$), while the relationship between binding integration and these relational rewards is insignificant. Binding integration is also significantly related to perceived role performance ($r = .62, p < .001$), a relational reward which was hypothesized to have a strong relationship to binding social ties. These results yield some support for Hypothesis 3a.

Furthermore, Table 1 also indicates that binding integration is significantly associated with the chronic strains ($r = .55, p < .001$), which indicates a strong relationship between social ties (at work and with family) and relational costs; nonbinding integration is unrelated to chronic strains. Thus, greater contact with others in family and work relationships is associated with higher levels of perceived stress in these relationships, as suggested by Hypothesis 3b.

Tables 2, 3, and 4 investigate whether these relationships hold when subjected to more rigorous analyses of ordinary least-squares regression (OLS). Table 2 presents two OLS equations which regress social integration on psychological distress, controlling for gender, marital status, education, income, age, and race. These data confirm the impressions given by the correlations, indicating support for Hypotheses 1 and 2. Equation 1 shows that individuals who have more social ties report less psychological distress ($b = -.14, p < .01$), as suggested by Hypothesis 1. Equation 2 presents the separate effects of binding and nonbinding integration on psychological distress. It indicates that binding integration is unrelated to psychological distress, while nonbinding ties are significantly and negatively associated with psychological distress ($b = -.15, p < .01$). This was predicted in Hypothesis 2. These findings suggest that *the direct beneficial effects of social integration, long noted in the literature* (Bell, LeRoy, and Stephenson 1982; Lin, Ensel, Simeone, and Kuo 1979; Miller and

Table 2. Standardized Regression Coefficients of
Social Integration on Psychological Distress ($N = 474$)

	Psychological Distress	
	Equation 1	Equation 2
All Integration	-0.14^{**}	
Binding Integration	—	-0.03
Nonbinding Integration	—	-0.15^{**}
Gender (F = 1)	0.19^{***}	0.19^{***}
Marital Status (Marr. = 1)	-0.04	-0.07
Employment (Yes = 1, No = 0)	-0.10^{*}	-0.12^{*}
Total Family Income	0.01	0.01
Education	-0.09	-0.08
Race (African Amer. = 1, Other = 0)	-0.10^{*}	-0.10^{*}
Age	-0.09	-0.06
Adjusted R^2	0.08	0.08

Notes: $^{*} p \leq 0.05$
       $^{**} p \leq 0.01$
     $^{***} p \leq 0.001$

Ingham 1979; Williams, Ware, and Donald 1981) *may be attributed to the nonbinding rather than the binding aspects of social ties.*

Tables 3 and 4 add relational costs and rewards to the OLS analyses. In Table 3, coefficients are presented for regressions of binding and nonbinding integration on relational costs and rewards. Here we find that binding integration is associated with higher levels of chronic strains ($b = .20, p < .001$), indicating a relationship between ties to family and work and relational costs, as suggested in Hypothesis 3b. Since this analysis controls for the number of roles associated with binding integration, this result indicates that the degree of contact within family and work domains is associated with higher levels of chronic strains or perceived stress within these roles. Regarding relational rewards from family and work, the level of binding integration is also positively related to perceived role performance. The relationships between binding integration and other relational rewards (perceived social support, mastery, and self-esteem) are not statistically significant. Thus, there is no relationship between binding integration and any relational reward other than perceived role performance. This effect yields further support for Hypothesis 3.

Also in Table 3, nonbinding integration is associated with increased levels of the relational rewards of self-esteem ($b = .11, p < .05$) and perceived social support ($b = .29, p < .001$), but is not significantly related to sense of mastery. The

Table 3.   Standardized Regression Coefficients of Relational Rewards on
Binding and Nonbinding Integration (All Analyses Control
for Gender, Marital Status, Age, Race, Education, Income,
Employment, and Number of Roles)

	Relational Cost	*Relational Rewards*			
	Chronic Strains	*Perceived Role Performance*	*Self-Esteem*	*Sense of Mastery*	*Perceived Social Support*
Binding Integration	0.20***	0.26***	0.05	0.03	0.03
Nonbinding Integration	−0.03	0.01	0.11*	−0.01	0.29***
Adjusted R^2	0.52	0.64	0.07	0.05	0.13

Notes: * $p \leq 0.05$
         ** $p \leq 0.01$
        *** $p \leq 0.001$

Table 4.   Standardized Regression Coefficients of Psychological Distress
on Relational Costs and Rewards (All Analyses Control for Gender,
Marital Status, Employment, Education, Income, Race, Age,
and Binding and Nonbinding Integration)

	Psychological Distress (Adjusted R^2 = 0.36)
Chronic Role Strains	0.20**
Perceived Role Performance	−0.11
Self-Esteem	−0.41***
Sense of Mastery	−0.10*
Perceived Social Support	−0.06

Notes: * $p \leq 0.05$
         ** $p \leq 0.01$
        *** $p \leq 0.001$

effects of nonbinding integration on chronic strains and perceived role performance are not statistically significant. Taken together these results indicate moderate support for Hypothesis 3a. In short, when a significant relationship was found between one index of integration and any relational cost or reward, the other form of integration had a nonsignificant effect on that variable.

In Table 4, coefficients are presented based on regressions of relational costs and rewards on psychological distress. We find here, as is commonly noted in the literature (Ensel and Lin 1991; Kessler and McLeod 1985; Pearlin and Lieberman 1979) that chronic strains, which are relational costs, are strongly associated with psychological distress (b = .20, p < .01). Moreover, as did previous studies

(Cohen and Wills 1985) we too find stress-buffering effects of the relational rewards of self-esteem ($b = -.41$, $p < .001$) and mastery ($b = -.10$, $p < .05$). However, other relational rewards—perceived social support and perceived role performance—do not exhibit these relationships. In addition, since the data were stratified we tested the full theoretical model for each of the four gender and marital status subgroups, as well as for the effects of gender and marital status, and did not find radically different or interesting results; moreover, the relatively small sample sizes made interpretation problematic. Taken together these data indicate some overall support for Hypothesis 4.

SUMMARY AND DISCUSSION

Researchers have argued that the relationships among social ties, social support and psychological well-being are in need of theoretical refinement (e.g., Sarason, Sarason, and Pierce 1990). We believe that in order to meaningfully assess the impact of social integration on mental health, both the nature of social ties and the costs or rewards associated with those ties must be examined. Accordingly, we argue that relationships with different responsibilities and obligations need to be considered separately. Furthermore, when these differences are taken into account, the differential costs and benefits of social ties will become clear. In other words, it is fruitful to break down social integration into two distinct concepts—binding and nonbinding integration—which differentially produce those costs and benefits that affect psychological well-being.

Our results indicated support for some aspects of this approach. We first replicated the reliable finding from the literature that social integration has moderate beneficial effects on psychological distress (Hypothesis 1). The data then also provided evidence consistent with our argument (Hypothesis 2) that this relationship can best be accounted for by relationships in which people more voluntarily engage, which we call nonbinding integration, but not at all by roles to which people feel bound for moral or financial reasons—that is, relationships with family and at work—which we call binding integration.

We also found some support for our third hypothesis, that binding integration is more strongly associated with relational costs and nonbinding integration with relational rewards. As predicted, we found that binding integration is associated with the relational cost of chronic strains; it was unrelated to any relational reward, with the exception of perceived role performance within family and work domains. Nonbinding integration is positively associated with two relational rewards (self-esteem and perceived social support) and unrelated to chronic strains—a relational cost. Taken as a whole, these findings indicate that binding and nonbinding integration have separate and distinct relationships to relational costs and rewards. With the exception of sense of mastery, binding and nonbinding integration exhibit complementary effects on costs and rewards.

The findings yielded more equivocal support for our claims regarding how relational costs and rewards mediate relationships between integration and mental health. Only one of the relational rewards—self-esteem—was related to nonbinding integration while also exhibiting a stress-buffering effect. (Another relational reward—perceived social support—had a stress-buffering effect but did not exhibit a relationship to any measure of integration.) This implies that the beneficial effects of nonbinding integration on psychological distress may be accounted for in part by the esteem-enhancing characteristics of these voluntary relationships. Perhaps it is because the relationships are voluntary not only for our respondents but also for those who befriend them that these relationships are more likely to yield these esteem effects. In addition, binding integration is associated with chronic strains, as predicted in Hypothesis 4. That is, ties to family and work are associated with higher levels of chronic strains, which, in turn, are associated with higher levels of psychological distress. Since our analyses control for the mere presence of the various work and family roles, these results indicate that a higher degree of contact with others is associated with greater perceived stress, which, in turn, is a significant predictor of mental health.

However, there are several problems inherent to these data which lead us to determine that our findings, however promising, are by no means conclusive. First, regarding the association between chronic strains and distress: while we are tempted to claim that the deleterious effect of binding integration on psychological distress can be accounted for in part by the strains emanating from work and family relationships, that may be unwarranted at this point. An alternative possibility is that the relationship between binding integration and chronic strains, as well as the lack of any such relationship between nonbinding integration and relational costs, may be an artifact of the manner in which chronic strains are typically measured. That is, all of the questions in this data set relating to chronic strains were identifiable with binding ties, but none were related to nonbinding ties. For example, respondents were asked about chronic strains related to such things as having children, relationships with their parents, or the work situation— ties that were also measured in the binding integration index. However, there were no chronic strains questions regarding relationships with neighbors and friends, or to church and club activities—that is, ties measured in the nonbinding integration scale. While this is a certainly a serious limitation to this study, at this point in time it may be unavoidable because *researchers, when constructing interview schedules, do not ordinarily ask about the relational costs from friendships, or club, or church associations* (indicators of nonbinding integration). However, researchers do regularly investigate (and find evidence of) the deleterious effects of family and work relationships (indicators of binding integration). Nevertheless, the apparent artifactual nature of these relationships precludes any conclusive interpretations. Thus one task of follow-up studies utilizing this approach would be to find or create databases which assess the relational costs of the nonbinding aspects of integration as well as the effects of these costs on psychological well-being. Similarly, the question of the artifactual nature of the relationship between relational costs and binding integration could also be minimized if future research could demonstrate a similar relationship between relational costs other than chronic strains (such as, for example, negative life events) and binding integration.

A related shortcoming of these data is that we were only able to measure a limited number of a theoretically infinite set of relational costs and rewards. For example, it could be argued that the list of possible relational rewards of binding integration was too small and should be increased. Indeed, the results might have been different had a more exhaustive set of rewards and costs been identified and measured. (Of course, we are inclined to believe that the results would have been more supportive of our claims.) For example, we would have liked to include a broad measure of perceived social support (including material and financial aid) as a relational reward. For relational costs, as mentioned, we would clearly have preferred to have had measures for the stresses of voluntary relationships, but we would also have liked to have been able to include a negative life events scale in the analysis to explore the possibility that binding relationships are more closely associated with such stressors than nonbinding relationships.

Finally, our finding that self-esteem was unrelated to the level of binding integration yet significantly associated with nonbinding integration is perhaps the most striking result of this study. We would like to conclude that the beneficial nature of nonbinding relationships is due in part to their "esteem-enhancing" effects (compare, e.g., Short, Sandler, and Roosa 1996). However, selection factors may be at play: individuals with high self-esteem may be more likely to seek out and to frequently engage in nonbinding relationships. Conversely, those with low self-esteem may be less likely to join recreational or religious clubs or groups or to take an active role in these organizations.

Despite these limitations, this study establishes some interesting new facts and provides some provocative conceptual leads for further exploration. First of all, as just discussed, we have explicitly taken up the charge of many observers to comparatively assess both the costs and benefits of being socially integrated. Our findings suggest that closer examination of the differential effects of these costs and benefits on mental health is a potentially fruitful avenue for further research.

Second, we have also presented a novel explanation for the common finding that global measures of social integration have moderately positive effects on psychological well-being. We have established that this is strongly related to those aspects of social integration which are voluntary (nonbinding) and unrelated to those aspects of integration which are not. This key finding could occasion myriad inquiries into *degrees of bindingness* of social roles and their effects on psychological outcomes. In other words, this analysis does not address variations in degree of "bindingness" or "nonbindingness" of relationships. All family or work ties are certainly not the same, and will vary, for example, in terms of degrees of obligation. Thus, the degree of "bindingness" or "voluntariness" of roles are matters for empirical investigation. Similarly, the degree to which time or energy expenditures are related to the commitment involved in binding relationships are matters for study. Unfortunately, these questions are rarely asked by researchers and the data used in this study do not address these important issues. While previous research provided us with some limited grounds to go beyond face validity and theorize along these lines, we hope this study might inspire future data gatherers to address this issue and develop measures that empirically assess what we have called the bindingness of relationships. This would allow us to learn (a) whether there are indeed clusters of "binding" and "nonbinding" roles, as we have surmised, and (b) whether they reliably predict relational costs, relational rewards, and psychological distress.

Moreover, it would be interesting to investigate whether the relative bindingness of relationships is simply a potent tool for objective analysis, or if it is subjectively perceived by people themselves. That is, although the binding/nonbinding distinction appears to have analytical power, further investigation could explore whether actors actually feel these distinctions. Some people may report that they find family relationships non-binding, while others may report they do not. Some may find relationships with friends or church associates to be involuntary and binding ties, while other may find these more voluntary. It would be interesting to see whether the relationships between forms of integration and psychological well-being map out more closely to objective measures of bindingness (as presented in this study) or to subjective measures. It would also be interesting to explore the extent to which any subjectively perceived bindingness is related to a contemporary cultural ambivalence to close relationships. David Karp, for example, in his study on depression, posits that contemporary Americans "are drawn to sources of connection for the comfort they provide, but equally fear what they perceive as the stifling features of all sorts of intimacies," adding that "many are afraid that supportive social bonds will evolve into bondage" (Karp 1996, p. 184).

Thus our study leads us to conclude (1) that there are both costs and rewards to social relationships than can and should be thoroughly analyzed, (2) that the degree to which people are structurally or culturally bound to relationships will help to determine the distribution of costs and rewards, and (3) in turn, that these differential effects of binding and nonbinding integration must be addressed in order to comprehend the impact of social ties on mental health. To this end, this study provides prospective researchers with a provocative conceptual scheme for investigating the psychological effects of social integration.

APPENDIX: CONTENTS OF SOCIAL INTEGRATION MEASURES

Binding Integration

Contact with Children

Score on Binding Integration Measure	Frequency of Contact with Children
0	Respondent has <u>no children</u>
1	Child lives <u>away from home</u> and respondent sees him/her <u>once a month or less</u>
2	Child lives away from home and respondent sees him/her at least <u>once a month</u>
3	Adult child (18 or over) lives with respondent or child lives away from home and respondent sees him/her <u>daily</u>
4	Child lives with respondent and is between the ages of 12–17
5	Child lives with respondent and is under the age of 12

Contact with Parents, In-Laws, and Relatives

Score on Binding Integration Measure	Frequency of Contact with Parents, In-Laws and the Relative with Whom the Respondent is in Contact with Most Often
0	Respondents parents, mother-in-law, and father-in-law are deceased
1	Several times a year or less
2	Several times a month or once a month
3	Once a week
4	Several times a week
5	Daily

Contact at Work

Score on Binding Integration Measure	Number of Hours per Week Spent Working with Others in Course of Respondent's Job
0	Respondent is not employed
1	Less than one hour
2	1–10
3	11–20
4	21–30
5	31 or more

Nonbinding Integration

Contact with Friends, Neighbors, Hobby or Leisure Activities, and Other Organizations/Clubs/Groups

Score on Nonbinding Integration	Frequency of Contact with Friends, Neighbors, Hobby or Leisure Groups, and Other Organizations/Clubs/Groups
0	Respondent does not have this social tie
1	Several times a year or less
2	Several times a month or once a month
3	Once a week
4	Several times a week
5	Daily

Contact at Church, Synagogue, or Other Religious Group

Score on Nonbinding Integration	Frequency of Contact with Religious Organization
0	Respondent does not have this social tie or only attends on religious holidays
1	Every few months
2	Once a month
3	Twice a month
4	Once a week
5	More than once a week

ACKNOWLEDGMENTS

A previous version of this paper was presented at the 1992 Annual Meeting of the American Sociological Association in Pittsburgh, Pennsylvania. We would like to thank Peggy Thoits, Brian Powell, Bernice Pescosolido, Mark G. Thompson, Kenneth Heller, and Pamela Braboy Jackson for comments on earlier drafts. This research was funded with grants from the National Institute of Mental Health (NIMH R01-MH43802), the Public Health Service (PHS S07-7031), and the Institute for Social Research at Indiana University. The authors are listed in alphabetical order, having made equal contributions to the work. Correspondence can be directed to Mitch Berbrier, Department of Sociology, University of Alabama in Huntsville, Huntsville, AL 35899. E-mail:berbrim@uah.edu.

REFERENCES

Barrera, M., Jr. 1986. "Distinctions Between Social Support Concepts, Measures, and Models." *American Journal of Community Psychology* 14: 413-445.

Bell, R.A., J.B. LeRoy, and J.J. Stephenson. 1982. "Evaluating the Mediating Effects of Social Supports upon Life Events and Depressive Symptoms." *Journal of Community Psychology* 10: 325-40.

Bellah, R., R. Masden, W. Sullivan, A. Swidler, and S. Tipton. 1985. *Habits of the Heart: Individualism and Commitment in American Life*. Berkeley, CA: University of California Press.

Bolger, N., A. DeLongis, R. Kessler and E. Wethington. 1989. "The Contagion of Stress across Multiple Roles." *Journal of Marriage and Family* 51(1): 175-183.

Bullers, S. 1997. "Social Support and Psychological Distress: The Mediating Role of Perceived Control." Paper presented at the Annual Meeting of the American Sociological Association, Toronto, August.

Cohen, S., and T.A. Wills. 1985. "Stress, Social Support, and the Buffering Hypothesis." *Psychological Bulletin* 98 (2): 310-57.

Coyne, J.C., and A. DeLongis. 1986. "Going Beyond Social Support: The Role of Relationships in Adaptation." *Journal of Consulting and Clinical Psychology* 54 (4): 454-460.

Cutrona, C.E. 1986. "Objective Determinants of Perceived Social Support." *Journal of Personality and Social Psychology* 50 (2): 349-355.

Dean, A., B. Kolody, and P. Wood. 1990. "Effects of Social Support from Various Sources on Depression in Elderly Persons." *Journal of Health and Social Behavior* 31: 148-161.

Derogatis, L.R., and P.M. Spencer. 1982. *The Brief Symptom Inventory (BSI): Administration, Scoring and Procedures, Manual I*. Baltimore, MD: Johns Hopkins University School of Medicine.

Eckenrode, J., and S. Gore (Eds.). 1990. *Stress Between Work and Family*. New York: Plenum Press.

Ensel, W.M., and N. Lin. 1991. "The Life Stress Paradigm and Psychological Distress." *Journal of Health and Social Behavior* 32: 321-341.

Fiore, J., J. Becker, and D.B. Coppel. 1983. "Social Network Interactions: A Buffer or a Stress?" *American Journal of Community Psychology* 11: 423-440.

Haines, V.A., and J.S. Hurlbert. 1992. "Network Range and Health." *Journal of Health and Social Behavior* 33: 254-266.

House, J.S. 1981. *Work Stress and Social Support*. Reading, MA: Addison-Wesley.

House, J.S., D. Umberson, and K.R. Landis. 1988. "Structures and Process of Social Support." *Annual Review of Sociology* 14: 293-318.

Hughes, M., and W. R. Gove. 1989. "Explaining the Negative Relationship Between Social Integration and Mental Health: The Case of Living Alone." Paper presented at the Annual Meeting of the American Sociological Association, San Francisco.

Jackson, P.B. 1992. "Specifying the Buffering Hypothesis: Support, Strain and Depression." *Social Psychology Quarterly* 55 (4): 363-378.

Karp, D.A. 1996. *Speaking of Sadness: Depression, Disconnection, and the Meanings of Illness*. New York: Oxford University Press.

Kessler, R.C., and M. Essex. 1982. "Marital Status and Depression: The Role of Coping Resources." *Social Forces* 61: 484-507.

Kessler, R.C., and J.D. McLeod. 1985. "Social Support and Mental Health in Community Samples." Pp. 219-240 in *Social Support and Health*, edited by S. Cohen and S.L. Syme. Orlando, FL: Academic Press.

Levinger, G., and L.R. Huesmann. 1980. "An 'Incremental Exchange' Perspective on the Pair Relationship: Interpersonal Reward and Level of Involvement." Pp. 165-188 in *Social Exchange: Advances in Theory and Research*, edited by K.J. Gergen, M.S. Greenberg, and R.H. Willis. New York: Plenum Press.

Lin, N., W.M. Ensel, R.S. Simeone, and W. Kuo. 1979. "Social Support, Stressful Life Events, and Illness: A Model and Empirical Test." *Journal of Health and Social Behavior* 20: 108-119.

Loscocco, K.A., and G. Spitze. 1990. "Working Conditions, Social Support, and the Well-Being of Female and Male Factory Workers." *Journal of Health and Social Behavior* 31: 313-327.

McLanahan, S., and J. Adams. 1987. "Parenthood and Psychological Well-Being." *Annual Review of Sociology* 13: 237-257.

Milardo, R.M. 1987. "Changes in Social Networks of Women and Men Following Divorce: A Review." *Journal of Family Issues* 8: 78-96.

Miller, P.M., and J.G. Ingham. 1979. "Reflections on the Life Events Link with Some Preliminary Findings." Pp. 313-336 in *Stress and Anxiety*, edited by I.G. Sarason and C.D. Spielberger. New York: Hemisphere.

Mirowsky, J., and C.E. Ross. 1986. "Social Patterns of Distress." *Annual Review of Sociology* 12: 23-45.

Morgan, D.L. 1989. "Adjusting to Widowhood: Do Social Networks Really Make It Easier?" *The Gerontologist* 29: 101-107.

Pagel, M.D., W.W. Erdly, and J. Becker. 1987. "We Get by With (and in Spite of) a Little Help from Our Friends." *Journal of Personality and Social Psychology* 53 (4):793-804.

Pearlin, L.I., E.G. Menaghan, M.A. Lieberman, and J.T. Mullan. 1981. "The Stress Process." *Journal of Health and Social Behavior* 22: 337-356.

Pearlin, L.I., and M.A. Lieberman. 1979. "Social Sources of Emotional Distress." Pp. 217-248 in *Research in Community and Mental Health: Volume 1*, edited by R. Simmons. Greenwich, CT: JAI Press.

Pearlin, L.I., and C. Schooler. 1978. "The Structure of Coping." *Journal of Health and Social Behavior* 19: 2-21.

Pescosolido, B., and S. Georgianna. 1989. "Durkheim, Suicide, and Religion: Toward a Network Theory of Suicide." *American Sociological Review* 54 (1): 1-39.

Procidano, M.E., and K. Heller. 1983. "Measures of Perceived Social Support from Friends and Family: Three Validation Studies." *American Journal of Community Psychology* 11: 1-24.

Rook, K.S. 1984. "The Negative Side of Social Interaction: Impact on Psychological Well-Being." *Journal of Personality and Social Psychology* 46 (5): 1097-1108.

Rook, K.S. 1992. "Detrimental Aspects of Social Relationships: Taking Stock of an Emerging Literature." Pp. 157-170 in *The Meaning and Measurement of Social Support*, edited by H.O.F. Veiel and U. Baumann. New York: Hemisphere Publishing Corporation.

Rosenberg, M. 1979. *Conceiving the Self*. New York: Basic Books.

Ross, C.E., J. Mirowsky, and J. Huber. 1983. "Dividing Work, Sharing Work, and In-Between: Marriage Patterns and Depression." *American Sociological Review* 48: 809-823.

Sarason, B.R., E.N. Shearin, G.R. Pierce, and I.G. Sarason. 1987. "Interrelations of Social Support Measures: Theoretical and Practical Implications." *Journal of Personality and Social Psychology* 52 (4): 813-832.

Sarason, I.G., B.R. Sarason, G.R. Pierce. 1990. "Social Support: The Search for Theory." *Journal of Social and Clinical Psychology* 9 (1): 133-147.

Schaefer, C., J.C. Coyne, and R.S. Lazarus. 1981. "The Health-Related Functions of Social Support." *Journal of Behavioral Medicine* 4: 381-406.

Short, J.L., I.L. Sandler, and M.W. Roosa. 1996. "Adolescents' Perceptions of Social Support." *Journal of Social and Clinical Psychology* 15: 397-416.

Simon, R.W. 1992. "Parental Role Strains, Salience of Parental Identity, and Gender Differences in Psychological Distress." *Journal of Health and Social Behavior* 33 (1): 25-35.

Steinberg, C. "The Influence of Positive and Negative Social Network Functions on Chronic Strain-Induced Psychological Distress in Male and Female Elderly Americans." Unpublished doctoral dissertation, University of Texas-Arlington.

Thoits, P.A. 1983. "Dimensions of Life Events that Influence Psychological Distress: An Evaluation and Synthesis of the Literature." Pp. 33-103 in *Psychosocial Stress: Trends in Theory and Research*, edited by H.B. Kaplan. New York: Academic Press.

Thoits, P.A. 1985. "Social Support and Psychological Well-Being: Theoretical Possibilities." Pp. 51-72 in *Social Support: Theory, Research, and Applications,* edited by I. G. Sarason and B. Sarason. The Hague, The Netherlands: Martinus Nijhof.

Thoits, P.A. 1991. "Identity Structures and Psychological Well-Being: Gender and Marital Status Considerations." Paper presented at the annual meeting of the Society for the Study of Social Problems, Cincinnati, Ohio.

Turner, R.J., C.F. Grindstaff, and N. Phillips. 1990. "Social Support and Outcome in Teenage Pregnancy." *Journal of Health and Social Behavior* 31: 43-57.

Turner, R.J., and S. Noh. 1983. "Class and Psychological Vulnerability Among Women: The Significance of Social Support and Personal Control." *Journal of Health and Social Behavior* 22: 357-367.

Umberson, D. 1987. "Family Status and Health Behaviors: Social Control as a Dimension of Social Integration." *Journal of Health and Social Behavior* 28: 306-319.

Umberson, D., and W.R. Gove. 1989. "Parenthood and Psychological Well-Being." *Journal of Family Issues* 10: 440-462.

Voight, J.D. 1996. "The Relation of Positive and Negative Social Network Characteristics to the Adjustment of Low-Income African-American Mothers." Unpublished doctoral dissertation, University of Illinois-Chicago.

Voydanoff, P., and B.W. Donnelly. 1989. "Work and Family Roles and Psychological Distress." *Journal of Marriage and Family* 51: 923-932.

Wellman, B., and S. Wortley. 1990. "Different Strokes from Different Folks: Community Ties and Social Support." *American Journal of Sociology* 96: 558-588.

Williams, A.W., J.E. Ware, Jr., and C.A. Donald. 1981. "A Model of Mental Health, Life Events, and Social Supports Applicable to General Populations." *Journal of Health and Social Behavior* 22: 324-336.

UNDERSTANDING DEPRESSION CARE IN THE HMO OUTPATIENT SETTING:
WHAT PREDICTS KEY EVENTS ON THE PATHWAY TO CARE?

Diana Shye, Donald K. Freeborn, and
John P. Mullooly

ABSTRACT

Depression is a major public health problem. The distress and functional and social disability it causes are costly to individuals and families, the health care system, and society. The majority of depressed patients are treated by primary care clinicians. Understanding is limited about the factors that affect the pathway to outpatient care for depression in HMO settings. This study describes, among members of a large U.S. health maintenance organization (HMO), the predictors of outcomes that represent progress on the pathway to care for depression, focusing in particular on the relative contribution of depressive symptom levels, gender, age, and other medical and nonmedical factors. The study population is an age/sex stratified sample of

Research in Community and Mental Health, Volume 11, pages 29-63.
Copyright © 2000 by JAI Press Inc.
All rights of reproduction in any form reserved.
ISBN: 0-7623-0671-8

HMO members aged 25+ ($N = 7,844$). Data sources include member survey questionnaires, medical charts, and automated utilization databases. Data were collected during a baseline year prior to the member's survey response date (1990–1992) and a follow-up year after that date. The study outcomes measured during the followup year were: study subjects' use of primary medical care; chart notations by a primary care clinician of depression diagnoses, antidepressant prescriptions, and referrals to specialty mental health care; and use of specialty mental health care. Predictor variables included age, gender, level of depressive symptoms, social role functioning, mental health care history, general health status, baseline health care utilization, sociodemographic characteristics, and relation to a personal primary care clinician (and the specialty of that clinician).

We found that primary care clinicians play the major role in caring for depression in this HMO. They recorded depression diagnoses for 4.8 percent of sample members and antidepressant prescriptions for the same proportion, identifying over 80 percent of all sample members who received any chart notation of these outcomes. In the absence of a baseline depression diagnosis, clinicians appeared to be responding mainly to current high levels of depressive symptoms. In the presence of a baseline depression diagnosis, age/gender interactions and sample members' relationship to a personal primary care clinician significantly predicted diagnosis and prescription but level of depressive symptoms did not. Having a family practice physician as personal primary care clinician was an important predictor of a depression diagnosis, an antidepressant prescription, and referral to and use of specialty mental health care. Our findings indicate that primary care clinicians are sensitive to manifestations of depression in their patients and that their diagnostic treatment and referral activities are appropriate responses to depressive symptom levels and past history of depression. Female gender appears to prompt the recognition and treatment of depression independent of symptom levels, especially among younger women. Recognition and treatment of depression in the primary care setting are likely to be enhanced when health care systems create structural and practice setting arrangements and processes that encourage continuity in patient–clinician relationships. More research is needed into the reasons for and the treatment cost and outcome consequences of differential therapeutic activism in depression care among general internists, family practice physicians, and other primary care clinicians. Future research should also explore the reasons for the independent effect of female gender in prompting the diagnosis and treatment of depression and study the attitudes and preferences of elderly people regarding depression care.

INTRODUCTION

As the most common psychiatric diagnosis at the community level, depression is increasingly recognized as a major public health problem (Robins and Regier 1991; Goldberg and Huxley, 1992; Culbertson 1997). Depressive disorders exact a heavy toll in functional disability, social morbidity, and health service utilization and costs, and the same is true of lower levels of depressive symptoms that do not meet DSM criteria (Broadhead et al. 1990; Johnson et al. 1992; Greenberg et al. 1993; Simon et al. 1995; Unutzer et al. 1997). Furthermore, depression is frequently a chronic, recurring condition (Klerman and Weissman 1992; Hays et al. 1995).

This study examines, among members of a large U.S. health maintenance organization (HMO), the predictors of outcomes that represent progress on the pathway to care for depression. It focuses on the relative contribution of depressive symptom levels, gender, age, relation to a personal primary care physician, and other medical and nonmedical factors. In particular, we sought to understand the factors that affect primary care clinicians' depression-related diagnostic and treatment decisions.

Depression and the Primary Care System

Primary care clinicians are regularly confronted with patients whose depressive symptoms run the gamut from transient and subclinical to chronic major depression. Research indicates that up to 50 percent of primary care patients suffer from depression, and most people who get treatment for depressive symptoms or disorders receive it solely in primary care settings (Goldberg and Huxley 1992). People with depressive disorders, as well as those with subclinical depressive symptoms, are especially high utilizers of primary care (Simon et al. 1995).

Recognition of primary care's role in depression care (Schurman et al. 1984) has stimulated efforts to better understand and to improve diagnosis and treatment in that setting. The U.S. Agency for Health Care Policy and Research (AHCPR) sponsored the development of a clinical practice guideline on treating depression in primary care (Depression Guideline Panel 1993), and subsequently supported research to evaluate strategies for implementing it in large group- and staff-model health maintenance organizations (Brown et al. 1995; Goldberg et al. 1998; Brown et al. 2000).

Studies comparing primary care clinicians' depression diagnostic and treatment decisions with specialty mental health clinicians' diagnostic criteria and treatment guidelines have repeatedly found the correspondence to be low. Primary care clinicians diagnose depression in patients who do not meet DSM criteria and often treat them with antidepressants. They also fail to diagnose and treat patients who do meet these criteria (Gerber et al. 1989; Perez-Stable et al. 1990; Pond et al. 1990; Goldberg and Huxley 1992). Furthermore, the care they provide to patients they identify as depressed frequently departs from mental health specialists' guidelines for antidepressant choice and dosage, duration of therapy, and follow-up care (Eisenberg 1992; Katon et al. 1992; Simon et al. 1993; Lin et al. 1995).

It is not clear, however, how these findings should be interpreted. Current depression care guidelines (Depression Guideline Panel 1993) are based on findings from clinical trials conducted mostly in specialty mental health care settings

(Gallo 1997; Schulberg et al. 1998). But primary care patients differ systematically from those typically seen in specialty mental health care settings: they generally have milder depressive symptoms, frequently express them in somatic form, often have little or no felt need for mental health care, and resist psychiatric labeling and treatment (Cleary 1989; Cleary and Burns 1990; Eisenberg 1992; Gerber et al. 1992; Johnson et al. 1992; Gotlieb et al. 1994; Simon et al. 1995; Docherty 1997). Research evidence on the appropriateness of specialty mental health depression guidelines for primary care is incomplete (Schulberg et al. 1998), but the differences between patients in the two settings have led commentators to suggest that comparing primary care clinicians' depression care diagnostic and treatment decisions with specialists' diagnostic criteria and clinical guidelines may not be helpful as a measure of the former's appropriateness (Goldberg 1992; Schwenk et al. 1996). Given the absence of depression guidelines with proven efficacy for the primary care setting and the prevalence of depressive symptoms among primary care patients, most experienced primary care clinicians have probably developed a pragmatic approach to depression care. At present, however, our understanding of the factors that prompt their diagnostic and treatment decisions is limited.

The Concept of a "Pathway to Care"

Goldberg and Huxley (1980, 1992) have described a pathway leading from mental illness in the community to mental illness in inpatient settings. They describe five "levels of care": (1) psychiatric morbidity in the community; (2) "hidden" (i.e., undiagnosed) psychiatric morbidity in the primary care system; (3) "conspicuous" (i.e., diagnosed) psychiatric morbidity in the primary care system; (4) psychiatric morbidity in the outpatient specialty mental health care setting; and (5) psychiatric morbidity in inpatient psychiatric care (Goldberg and Huxley 1980, 1992).

Between any two levels is a "filter" through which individuals must pass to reach the next level of care. The first filter, between the community and the primary care setting, is individuals' culturally conditioned illness behavior. Progress from the second to higher levels of care is largely determined by primary care and specialist clinicians' decisions (i.e., the primary care clinician's skill in diagnosing depression [the second filter]; the primary care clinician's propensity to refer the patient to specialized mental health services [the third filter]; and the mental health specialist's decision to admit the patient to inpatient care [the fourth filter] [Goldberg and Huxley 1992]). This model was developed to describe health care systems characterized by mandatory universal health care coverage, in which primary care clinicians act as gatekeepers to specialty mental health services and progress to higher levels of specialty care is coordinated regionally.

It may, however, be useful to broaden Goldberg and Huxley's model to take into account the interactions between patients' and clinicians' decisions and actions. Individual illness behavior may determine initial entry into the primary care system in an episode of depression, but return visits for the same episode are a function of both patient and clinician behavior. Likewise, clinicians' ability and propensity to diagnose and treat depression and to refer patients to specialty care are affected by the way patients present their symptoms and by patients' attitudes toward psychiatric labeling and specialty mental health treatment (Goldberg 1982, 1990). Adapting the model in this way also makes it more useful for describing health care systems in which people can access outpatient specialty mental health services directly from the community on their own initiative.

Influences on the Pathway to Care

Severity of depressive symptoms, level of functional disability, and prior mental health history understandably influence the way people move along the pathway described by Goldberg and Huxley (1992). But research indicates that some nonmedical factors also have important effects—in particular, patients' gender and age. Why this should be so is unclear, but it raises questions about the appropriateness of clinicians' depression diagnostic and treatment decisions.

The Role of Gender

Women are more likely than men to pass through the first filter of individual illness behavior and access the primary care system (Goldberg and Huxley 1980; Strickland 1988; Culbertson 1997; Sprock and Yoder 1997). Once they do so, they are also more likely to present psychological complaints and to express symptoms congruent with a DSM diagnosis of depression (Angst and Dobler-Mikola 1984; National Institutes of Mental Health D/ART Program 1987; Page and Benesch 1993). This may be why they also pass through the second filter—diagnosis by a primary care clinician—more readily than men (Goldberg and Huxley 1980). Possible explanations for these findings include a higher true incidence of depression among women due to biological factors and social influences and/or a gender bias in current depression diagnostic criteria that creates a self-reinforcing tendency for clinicians to recognize and diagnose depression in women and overlook it in men (Angst and Dobler-Mikola 1984; Sprock and Yoder 1997). Some research shows that primary care clinicians may more frequently refer men to specialty mental health care than women (Goldberg and Huxley 1982, 1992).

The Role of Age

Major depression appears to be less prevalent in the elderly than in younger people (Robins and Regier 1991; Depression Guideline Panel 1993; Culbertson 1997), but the reasons for this are not well understood. Elderly people may be less likely than younger people to express their distress in "mental" or psychological terms (Veroff et al. 1981). Alternatively, the psycho-physiological presentation of depression may differ in the elderly (Newman et al. 1991)—in which case, current DSM categories may have limited value for diagnosing depression in the elderly (Newman et al. 1991; Culbertson 1997). But subclinical depression—sometimes chronic in nature—is quite prevalent among the elderly and associated with significant morbidity and health care costs (Blazer 1991; Callahan et al. 1994; Unutzer et al. 1997). Whatever the reason, depressed elderly seeking care often present nonspecific somatic and psychological complaints, which may reduce the likelihood of a diagnosis. In addition, differential diagnosis is more complex in the elderly: depression may be a side effect of medications or a symptom of some diseases (Blazer 1991). Elderly patients are also less likely than younger ones to pass through the third filter of referral to specialty mental health services—perhaps because they have less trust than younger people in mental health specialists (Waxman et al. 1984; Goldstrom et al. 1987; Goldberg and Huxley 1992).

Other Influences on Movement along the Pathway to Care

Poor self-perceived health status and the presence of chronic illness increase the likelihood of crossing the first filter into the primary care system for people with all levels of psychological distress (Goldberg and Huxley 1992). Patient factors found to increase clinician identification of depression include being unmarried, being unemployed, having low education, and high medical care utilization (Marks et al. 1979; Boardman 1987). Some clinician characteristics, such as being older, more experienced, more settled in practice, and having more interest in the psychosocial aspects of care appear to be associated with greater recognition of mental distress (Goldberg and Huxley 1992; Parchman 1992). The interaction between a somaticizing patient and a time-pressured primary care clinician with a "low bias" toward recognition of psychological distress may result in a collusion between them to avoid dealing with depression (Goldberg 1990). The fourth filter of referral to inpatient care appears to be mainly a function of the severity of the patient's psychiatric disorder (Goldberg and Huxley 1992).

Objectives of the Present Study

The purpose of this study was to contribute to understanding of the factors that affect primary care clinicians' depression-related diagnostic and treatment decisions. We undertook to identify, among members of a large U.S. health maintenance organization (HMO), the predictors of patient outcomes that represent progress on the pathway to outpatient care for depression: use of primary care; chart notation of a depression diagnosis, antidepressant drug prescription or referral to specialty mental health care by a primary care clinician; and use of specialty mental health care. Using hierarchical multivariate logistic regression models to reveal direct and indirect effects, we elucidated the relative contributions of depressive symptoms, patient gender and age, social role impairment due to emotional problems, mental health care history, general health status, socioeconomic characteristics, and relationship to a personal primary care clinician.

METHODS

Study Site

The study site was Kaiser Permanente Northwest Region (KPNW), an established prepaid group practice HMO that provides comprehensive care to over 420,000 members in the Portland, Oregon–Vancouver, Washington metropolitan area. Members resemble the overall area population in age distribution as well as health status and sociodemographic characteristics (Greenlick et al. 1988; Freeborn and Pope 1994). Primary care is provided by internal medicine and family practice physicians, physician assistants, and nurse practitioners. Much care for depression and anxiety, as well as for more serious mental illnesses, is provided by primary care clinicians (Johnson and McFarland 1994; McFarland et al. 1996). Mental health care received in the primary care setting is covered under members' medical benefits, rather than by their mental health benefits. Ninety-six percent have mental health care benefits that vary depending on the member's employer group, but are generally as good or better than those offered by other U.S. HMOs (Levin 1990). HMO members may self-refer to the HMO's specialty mental health department or be referred by their primary care clinicians. In 1992, 3.4 percent of the HMO's members made at least one outpatient visit to a specialty mental health clinician.

Study Subjects

The study subjects were an age/sex stratified sample of HMO members aged 25 and older who responded in 1990–1992 to the HMO's ongoing Current Member Survey (CMS) conducted by the Center for Health Research. All were HMO members for at least 1 year prior to and following their survey response date. The overall CMS response rate was about 60 percent; older persons and women had higher response rates than younger people and men. Once age and gender were controlled, respondents and non-respondents were comparable on depression-related variables measured in the year surrounding their survey response date (e.g., rate of and average number of specialty mental health visits, rate of and average number of antidepressant dispenses). We oversampled men respondents to the CMS in order to increase their proportion among sample subjects with outcomes of interest. Table 1 shows the age/sex distribution of the study sample, and the sampling fractions of the CMS for each age/sex strata.

Table 1.   Age/Sex Distribution of Study Sample and
Current Member Survey Respondent Sampling Fractions

	Age 25-54		Age 55-69		Age 70+		Total	
	N	CMS sampling fraction	N	CMS sampling fraction	N	CMS sampling fraction	N	CMS sampling fraction
Male	1594	.58	1571	1.00	986	1.00	4151	.79
Female	1197	.32	1221	.68	1275	1.00	3693	.54
Total	2791	.43	2792	.83	2261	1.00	7844	.65

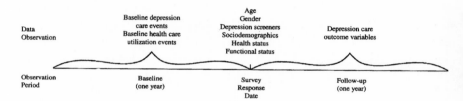

Figure 1.   Study Framework and Timing of Data Observations

Study Design

Each sample member had a two-year observation period, centered on his or her CMS response date (the "index time point"). The year prior to survey response was the "baseline year" and the year following it was the "follow-up year." Since the CMS was an ongoing survey, the overall study time frame for the 1990-1992 CMS respondents extended from January 1, 1989, to December 31, 1993. Figure 1 shows the timing of data observations for sample members.

Data Sources

Study data came from three principal sources:

1. *The HMO Current Membership Survey (CMS).* This self-administered survey provided data on sociodemographic characteristics, depressive symptom levels, health status and limitations on functional status and social activities (as measured by SF-36 scale scores [Ware and Sherbourne 1992]).

2. *Outpatient medical records.* Information about every medical care contact is recorded in a single outpatient medical record for each HMO member. Trained medical record technicians abstracted from sample members'

medical records all depression and other psychiatric diagnoses, antidepressant drug orders, and referrals to specialty mental health care noted by any HMO clinician during the baseline and follow-up years.

3. *HMO automated administrative databases.* A unique health record number for each HMO member links all patient-related data from these databases, while a unique number for each clinician identifies care given by him or her. The HMO's appointment and encounter system provided the data on sample members' utilization of primary care, nonmental health specialty care, and specialty mental health care. The outpatient pharmacy utilization system yielded data on dispenses of antidepressants and other medications and was used in computing sample members' Chronic Disease Scores (Johnson et al., 1994). The HMO's member information system provided data on mental health benefit and some information used to assign sample members to personal primary care clinicians. The hospital admissions, discharge, and transfer system yielded data on sample members' general inpatient stays. An automated system for processing claims for covered services provided by non-HMO providers was the source of data on psychiatric hospitalizations and on outpatient visits to non-HMO mental health clinicians.

Study Measures

Outcome variables including five dichotomous outcomes, measured in the follow-up year, reflect key events along the pathway from the first level of care for depression (the community) to the fourth level (use of specialty outpatient mental health services) (Goldberg and Huxley 1992).

1. *Use of outpatient primary medical care* (passage through the first filter);
2. *Primary care clinician's medical chart diagnosis of depression* (passage through the second filter);
3. *Primary care clinician's medical chart notation of an antidepressant drug prescription* (reaching the third level of care);
4. *Primary care clinician's medical chart notation of a referral to specialty mental health care* (passage through the third filter);
5. *Use of specialty mental health care* (reaching the fourth level of care).

Independent variables included three variables previously shown to be strongly related to the outcomes under study:

1. *Patient age group* (25-54, 55-69, 70+);
2. *Gender* (male, female);
3. *Level of depressive symptoms.* We developed an ordinal measure based on sample members' responses to two previously validated short depression screening scales (measured at the index timepoint): (a) the *Brief Screening*

Instrument (BSI-8), an eight-item scale that describes mood and neuroveg-etative symptoms and screens for dysthymia by asking about duration of depressed mood (Burnam et al. 1988); (b) the *Mental Health Inventory (MHI-5)* a scale containing five items, different from those on the BSI-8, that screen for symptoms of depressed mood experienced "in the last four weeks" (Berwick et al. 1991). We used the accepted cut-points of of these scales to define a positive depression screening status.

The correlation between BSI-8 and the BHI-5 summary scores in our sample was high ($r = .68$), but only about 40 percent of those who screened positive for depression on *one* of them screened positive on *both*. The highest correlation between individual items of the two scales was 0.59 (for the BSI-8 item "feeling downhearted and blue" with the MHI-5 item "felt depressed"). Correlations between most other items were considerably lower. Thus, the BSI-8 and the MHI-5 appeared to be complementary: each characterized depression-related symptoms in a subset of individuals who would not have been identified by the other. We therefore used sample members' MHI-5 and BSI-8 screening status to construct an ordinal measure of level of depressive symptoms (neither screener positive = low, one screener positive = moderate, both screeners positive = high).

Other predictor variables included measures of sample members' sociodemo-graphic characteristics, health and functional status, prior general and mental health care utilization and relationship to a "personal primary care clinician" as possible confounding and intervening factors in the relationships between the independent variables and study outcomes. The Appendix contains details about all study variables.

The Chronic Disease Score (CDS) uses automated pharmacy utilization data to compute a measure of chronic disease burden. Developed by Von Korff and col-leagues (1992) and later replicated and validated by Johnson and colleagues (1994) among HMO members in the present study site, it has good reliability, construct validity with the RAND-36 instrument (Ware and Sherbourne 1992) and predictive validity for health care visits and hospitalizations (Johnson et al. 1994; Hornbrook et al. 1996). An advantage of the CDS for the present study is its demonstrated lack of association with depression and anxiety (Johnson et al. 1994). Sample members' CDS scores were based on pharmacy utilization data for the baseline year.

We determined sample members' relationship to a personal primary care clini-cian in one of two ways. Those the HMO's automated member information system listed as having chosen a personal clinician were assigned to that clinician if they had made at least one visit to him or her during their two-year observation period. Among those with *no* personal clinician, we identified from the HMO's appoint-ment and encounter system people who had made office visits to primary care cli-nicians during their two-year observation period and applied a previously validated four-step algorithm (Johnson et al. 1994) to assign them to a "personal clinician"

Table 2. Distribution of Sample Members by Method of
Assignment to Personal Primary Care Clinicians

Source of Assignment	Criterion	Has "Personal Primary Care Clinician"	N	%
HMO's member information and appointment and encounter systems	Sample member had at least one primary care visit during study observation period with a clinician designated as his/her personal primary care clinician.	Yes	4758	60.7
Algorithm (Step 1)	All primary care visits to one clinician.	Yes	873	11.1
Algorithm (Step 2)	Primary care visits to two clinicians, 51% to a single clinician.	Yes	281	3.6
Algorithm (Step 3)	Primary care visits to three clinicians, 51% to a single clinician.	Yes	252	3.2
Algorithm (Step 4)	Primary care visits with multiple clinicians, less than 51% to a single clinician.	No	715	9.1
HMO's Member Information and Appointment and Encounter systems	No designated personal primary care clinician; no primary care visits during study observation period.	Unassignable	965	12.3
		Total	7844	100

based on the identity of the clinicians they had visited. The 965 sample members who made *no* office visits to *any* primary care clinician during the two-year observation period were thus "unassignable" to a primary care clinician. They were significantly younger than the "assignable," more likely to be men and less likely to be depressed or to have chronic health problems. This created some confounding between lack of a personal primary care clinician and a negative value on the first study outcome variable (use of primary care in the follow-up year), and we took this into account in our data analyses. Table 2 describes the study population according to method of assignment to a personal primary care clinician.

Statistical Analyses

In order to elucidate direct and indirect relationships between predictor and outcome variables, we developed a single hierarchical multivariate logistic regression model for each outcome, based on preliminary bivariate and multivariate logistic regression analyses. Each model included significant predictor variables from preliminary logistic regression models (after necessary exclusions to prevent multicolinearity). For the second and third outcomes (chart notation of a depression diagnosis and chart notation of an antidepressant prescription), the predictive models were very different for sample members who *had* and *had not* received a medical chart notation of a diagnosis in the baseline year, so we created separate hierarchical models for these two groups.

The structure of all the hierarchical models reflected the same underlying conceptual model. We assumed that self-perceived limitations in social role functioning would be the most fundamental factor underlying our study outcomes (Suchman 1965). Therefore, the sample member's relevant SF-36 scale score entered the model at Step 1. Given previous findings of age and gender differences in our outcomes and primary care patients' tendency to somaticize depression, we assumed that age/gender interaction effects would take precedence over the effects of depressive symptoms; thus these interactions entered at Step 2. We also assumed that depressive symptoms and other mental-health-related measures would predominate over general health status in predicting the outcomes under study and that more recent depression experience would take precedence over more distant experience. Thus, Step 3 entered depressive symptom levels measured at the index time point, Step 4 entered mental health care history, and Step 5 entered general health status. To test for a possible effect of recent health care use independent of health status (Freeborn et al. 1990; Goodman et al. 1991), Step 6 introduced baseline year health care utilization. Socioeconomic status is largely unrelated to access to care in this HMO (Freeborn and Pope 1994), so a compound variable based on income level and social class entered at Step 7 along with marital status. Sample members' relationship (or lack thereof) with a personal primary care clinician entered at Step 8 (except for models that included sample members "unassignable" to a personal primary care clinician).

We tested the fit of the models to the data using the Hosmer-Lemeshow goodness-of-fit statistics and the concordance between the observed outcomes and those predicted by the model using the Goodman-Kruskal's gamma. Because the number of positive events for most outcomes was small, the resulting multivariate models were overfitted. The increase in -2 log likelihood between the full models and trimmed models was not statistically significant for any outcome; thus we present the full models here. For the sake of parsimony, only the final step of the (untrimmed) models (including odds ratios [O.R.], p-values for individual predictors and maximum-rescaled R^2 for the model) is shown in the tables; earlier steps are described in the text. The full hierarchical logistic regression models for each outcome can be viewed at the following web site:http://www.kpchr.org/info/present/ depcare-shye.pdf.

STUDY FINDINGS

Preliminary Analyses

The univariate distribution of all study variables is shown in the Appendix. Only 4.3 percent of sample members had a high level of depressive symptoms;

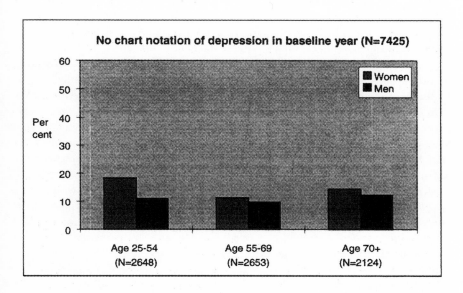

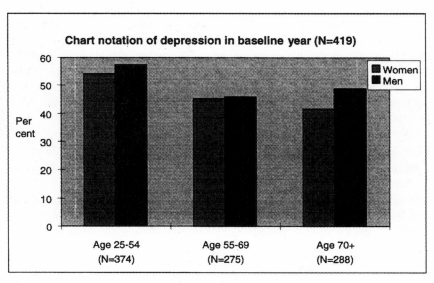

Figure 2. Proportion of Sample Numbers with
Significant Depressive Symptoms
(Depressed on MHI-5, BSI-8 or Both)

85.5 percent had low symptoms. Depressive symptom level was correlated with measures of health and functional status—especially with measures specific to mental health (e.g., the SF-36 scale for the degree of social role limitations due to emotional problems [phi = .39] and sample members' self-reports of ever having been told by a doctor they have depression [phi = .36]). Figure 2 shows the proportion of men and women in different age groups who screened positive on at least one of the two depression screening instruments. Among sample members with *no* baseline year chart diagnosis of depression, this proportion ranged from about 10-20 percent and women's rates were significantly higher than men's only among those aged 25-54. Among sample members *with* a baseline year depression diagnosis, the proportion who screened positive was several times higher and did not differ significantly by age or gender.

Most sample members (77.5 percent) used primary care in the follow-up year, but other study outcomes were far rarer. Only 4.8 percent received a primary care clinician's depression diagnosis, and an identical proportion received an antidepressant prescription. Nearly three percent made at least one outpatient specialty mental health visit—higher than the 1.2 percent who were referred by primary care clinicians for such care. (This reflects the fact that members of this HMO can self-refer to the fourth level of care.) A primary care depression diagnosis and antidepressant prescription were strongly but by no means fully correlated (phi = .59, p = .001). This correlation was stronger among people with a baseline year chart diagnosis of depression: in this group, 92.4 percent of those who received a prescription also received a diagnosis, and 72.4 percent of those who received a diagnosis also received a prescription. Among those with *no* baseline year depression diagnosis, the analogous proportions were 40.9 and 49.0 percent. Diagnosis and referral were moderately correlated (phi = .30, p = .001) and other study outcomes were only weakly associated.

Table 3 below shows data abstracted from sample members' medical records about depression diagnoses, antidepressant prescriptions, and referrals to specialty mental health care made by *any* of the HMO's clinicians. Over 80 percent of sample members with at least one follow-up year chart notation of any of these outcomes received them from primary care clinicians (though not necessarily *only* from them). Specialist clinicians—whether mental health or nonmental health—made such chart notations for less than a fifth of sample members who received them at all.

Hierarchical Multivariate Models

P-values for the Hosmer-Lemeshow goodness of fit statistics ranged from .21 to .84, indicating that all the models fit the data adequately. Goodman-Kruskal's gammas for the models were in the middle range (0.51 to 0.67), reflecting moderate concordance between observed and predicted outcomes. The exception was a

Table 3. Distribution of Sample Members with Specific Follow-up Year Medical Chart Notations, by Type of Clinician Making the Notation

Chart notation made by:	Sample members with Depression diagnosis		Sample members with antidepressant prescription		Sample members with referral to mental health department	
	N	% of total with any Dx	N	% of total with any Rx	N	% of total with any referral
Any KPNW clinician	458	100.0	444	100.0	118	100.0
Primary care clinician	377	82.3	375	84.5	98	83.1
Non-mental health specialist clinician	74	16.2	86	19.4	22	18.6
Specialty mental health clinician	87	19.0	69	15.5	NA	NA

gamma of 0.19 for the model predicting follow-up year use of mental health care (see below).

Use of Outpatient Primary Medical Care in the Follow-up Year

As expected, the primary predictors of this outcome were age, gender, and general health status. (Results not shown. The full model can be viewed at the following website: http://www.kpchr.org/info/present/ depcare-shye.pdf.) Utilization increased with age in both sexes and was higher among women at all ages, as well as among those with chronic illness and a recent deterioration in health. Level of self-reported depressive symptoms did not affect use of primary care, nor did mental health care history. At Step 1, poor social role functioning (because of physical or emotional problems) strongly predicted primary care use. But controlling for other factors showed that this was because of higher social role impairment among people with prior mental health care, chronic disease, and a recent deterioration in general health status. This model included sample members who were unassignable to a personal primary care clinician, so Step 8 (relationship to a personal clinician) was not used. The maximum-rescaled R^2 for the model was .28, indicating that the predictor variables accounted for about 28 percent of the variance in this outcome (Nagelkerke 1991).

Primary Care Clinicians' Diagnosis and Antidepressant Treatment of Depression:
Models for sample members with no baseline chart diagnosis of depression

Table 4 shows the final steps in the hierarchical logistic regression models for a primary care clinician's depression diagnosis and antidepressant prescription.

Primary care clinician's depression diagnosis. The sample member's self-report of ever having been "told by a doctor you have depression" was the strongest predictor of this outcome in this subgroup ($O.R. = 4.7$). Having a family practice or internal physician as personal clinician ranked next ($O.R. = 3.36$ and 3.31, respectively), followed by screening depressed on both screeners ($O.R. = 2.5$) and being a woman in the 25-54 age range ($O.R. = 2.5$).

Poor social role functioning due to physical or emotional problems affected the likelihood of a depression diagnosis directly and indirectly (through the more impaired social functioning of people with high levels of depressive symptoms, prior mental health care history, chronic disease and recent deterioration in health status). A number of significant age/gender interaction effects emerged initially, but once other factors were controlled, only one age/gender interaction term— women aged 25-54—retained a significant direct effect on the likelihood of diagnosis (Step 8, $O.R. = 2.5$).

Depressive symptom levels affected the likelihood of a depression diagnosis both directly and indirectly. Before other factors were controlled (Step 3), people with moderate symptom levels appeared 2.4 times more likely than those with low symptoms to be diagnosed as depressed, and the odds ratio for those with high symptom levels was double that ($O.R. = 5.0$). Once mental health care history entered the model (Step 4), these odds ratios dropped to 1.7 and 2.9, respectively, because people with a history of mental health care had higher depressive symptom levels.

Self-reported recent deterioration in general health status also significantly increased the likelihood of a depression diagnosis, even after other factors were controlled ($O.R. = 1.7$). The explanatory power of the variables in the model was modest (maximum-rescaled $R^2 = .18$).

Antidepressant prescription. A baseline year antidepressant prescription was the strongest predictor of a follow-up year prescription among people with no baseline year chart diagnosis of depression ($O.R. = 2.37$). Following in relative importance were the age/gender interactions terms that described women under 70 ($O.R. = 2.33$ and 2.21 for women aged 25-54 and 55-69, respectively), high social role impairment ($O.R. = 2.16$), and having a family practitioner as personal clinician ($O.R. = 2.01$). Poor social role functioning affected antidepressant prescribing directly ($O.R. = 2.17$) and indirectly (because people with higher depressive symptoms had more impaired social functioning [Step 3], as did those with a mental health care history [Step 4] and with poor general health status [Step 5]).

Table 4. Predictors of a Follow-up Year Medical Chart Notation of Depression Diagnosis and Antidepressant Prescription by a Primary Care Clinician: *Sample Members with <u>No</u> Baseline Year Notation of Depression*

	Final Step of Model (Step 8)			
	Chart Notation of Depression Diagnosis $N = 5695$; Number of events $= 188$		Chart Notation of Anti-depressant Prescription $N = 5695$; Number of events $= 220$	
Variables Entered:	O.R.	P	O.R.	P
Step 1: Social Role Functioning				
High limitations due to physical or emotional problems	1.74	.003	2.16	.0001
Step 2: Age x Gender				
Men Age 25-54 (Reference)	—	—	—	—
Women Age 25-54	2.52	.002	2.33	.009
Men Age 55-69	0.64	.18	1.07	.85
Women Age 55-69	1.32	.37	2.21	.01
Men Age 70+	0.64	.21	1.18	.64
Women Age 70+	1.13	.69	1.70	.10
Step 3: Level of Depressive Symptoms				
Not depressed on MHI-5 or BSI-8 (reference)	—	—	—	—
Depressed on either MHI-5 or BSI-8	1.72	.01	1.16	.49
Depressed on both MHI-5 and BSI-8	2.95	.0001	1.88	.02
Step 4: Mental Health Care History				
Ever told by Dr. you have depression? (Yes)	4.74	.0001	1.88	.004
Baseline Dx: anxiety, sleep/family problems	1.52	.07	2.07	.0003
Baseline antidepressant prescription	1.05	.72	2.37	.0001
Baseline referral to specialty mental health	2.08	.13	0.85	.80
Baseline specialty mental health visit	0.37	.08	0.18	.02
Step 5: General Health Status				
Chronic Disease Score > 0 (any chronic disease)	1.32	.15	1.17	.39
Health now worse than last year? (Yes)	1.74	.003	1.26	.18
Step 6: Baseline Health Care Utilization				
High (4+) baseline year primary care visits	1.29	.14	1.40	.04
Step 7: Sociodemographic Characteristics				
Unmarried	1.01	.96	1.10	.57
Socioeconomic status				
Middle/upper class (reference)	—	—	—	—
Lower/working class	0.93	.66	1.01	.95
Step 8: Relation to Primary Care Clinician				
None (reference)	—	—	—	—
Internal medicine physician	3.31	.0001	1.59	.06
Family practice physician	3.36	.0001	2.01	.007
Physician assistant	2.08	.22	0.85	.80
Nurse practitioner	1.70	.37	1.10	.86
Maximum-rescaled R^2	.18		.17	

Women of all ages were about twice as likely as men to get a prescription, but this gender difference decreased with increasing age. A high level of depressive symptoms significantly increased the likelihood of an antidepressant prescription, as did having a baseline year chart notation of a mild non-depression psychiatric diagnosis (such as anxiety, sleep disturbances, or family problems) (*O.R.* = 2.07) and having "ever been told by a doctor you have depression" (*O.R.* = 1.88). Patients of family practitioners and internal medicine physicians were significantly more likely than those with no personal clinician or with allied health clinicians to receive a prescription. The explanatory power of the variables in this model was modest ($R^2 = 0.17$).

*Models for sample members **with** a baseline chart diagnosis of depression.*

Table 5 shows the hierarchical logistic regression models for a depression diagnosis and an antidepressant prescription for these sample members.

Primary care clinician's diagnosis of depression.   Of the independent variables we studied, only age had a significant direct effect on the likelihood of diagnosis in this subgroup. People aged 55 and older were about three times more likely than younger people to be diagnosed as depressed—regardless of gender. This effect became increasingly clear as additional variables were controlled. Baseline year antidepressant treatment and self-report of a previous depression diagnosis significantly *increased* the likelihood of a diagnosis (*O.R.* = 2.37 and 1.88, respectively), while baseline year specialty mental health care *reduced* it dramatically (*O.R.* = 0.25). People whose personal primary care clinician was a family practice physician were much more likely to be diagnosed at follow-up than others (*O.R.* = 4.0 relative to those with no primary care clinician) and internal medicine physicians' patients also had significantly increased diagnosis rates (*O.R.* = 2.69). This model had moderate explanatory power ($R^2 = .28$).

Antidepressant prescription.   The age/gender interaction terms that describe women aged 55 and older were the strongest predictors of this outcome among sample members who had been diagnosed as depressed in the baseline year, and this effect increased as additional variables were controlled. Men aged 55+ were also marginally more likely than younger people to receive a prescription. Next in importance was having a family practice physician as personal clinician rather than another type of clinician or no personal clinician (Step 8, *O.R.* = 2.96).

In contrast with the lack of effect of depressive symptom levels on depression diagnosis in this subgroup, such symptoms did strongly affect the likelihood of an antidepressant prescription (Step 8, *O.R.* = 2.09). Three mental health care history variables also had significant effects: "ever having been told by a doctor" that one is depressed and baseline year antidepressant treatment increased the chance of a primary care clinician's prescription at follow-up (*O.R.* = 1.99 and

Table 5. Predictors of a Follow-up Year Medical Chart Notation of Depression Diagnosis and Antidepressant Prescription by a Primary Care Clinician: *Sample Members* **with** *a Baseline year Notation of Depression*

| | Final Step of Model (Step 8) | | | |
| | Chart Notation of Depression Diagnosis N = 384; Number of events = 179 | | Chart Notation of Anti-depressant Prescription N = 384; Number of events = 141 | |
Variables Entered:	O.R.	P	O.R.	P
Step 1: Social Role Functioning				
High limitations due to physical or emotional problems	1.40	.27	1.19	.60
Step 2: Age x Gender				
Men Age 25-54 (Reference)	—	—	—	—
Women Age 25-54	0.87	.77	1.19	.75
Men Age 55-69	2.70	.07	2.34	.16
Women Age 55-69	2.97	.03	4.33	.007
Men Age 70+	3.14	.04	2.31	.17
Women Age 70+	3.31	.02	4.50	.007
Step 3: Level of Depressive Symptoms				
Not depressed on MHI-5 or BSI-8 (reference)	—	—	—	—
Depressed on either MHI-5 or BSI-8	0.86	.61	1.15	.65
Depressed on both MHI-5 and BSI-8	1.42	.33	2.39	.02
Step 4: Mental Health Care History				
Ever told by Dr. you have depression? (Yes)	1.74	.04	1.99	.01
Baseline diagnosis of anxiety, sleep/family problems	0.73	.21	1.04	.89
Baseline antidepressant prescription	1.29	.0001	1.34	.0001
Baseline referral to specialty mental health	1.35	.15	0.89	.54
Baseline specialty mental health visit	0.25	.0001	0.13	.0001
Step 5: General Health Status				
Chronic Disease Score > 0 (any chronic disease)	0.62	.11	0.68	.22
Self-reported health status = Poor (SF-36)	1.88	.03	1.68	.08
Health now worse than last year? (Yes)	0.80	.45	NA	NA
Step 6: Baseline Health Care Utilization				
High (4+) baseline year primary care visits	0.62	.07	0.78	.36
Any baseline Urgency Care/ER visit (yes)	1.15	.05	NA	NA
Step 7: Sociodemographic Characteristics				
Unmarried	0.73	.23	0.52	.02
Socioeconomic status				
Middle/upper class (reference)	—	—	—	—
Lower/working class	0.91	.72	1.10	.74
Step 8: Relation to Primary Care Clinician				
None (reference)	—	—	—	—
Internal medicine physician	2.69	.01	1.36	.46
Family practice physician	3.99	.002	2.96	.01
Physician assistant	0.79	.85	0.68	.77
Nurse practitioner	2.21	.32	1.79	.47
Maximum-rescaled R^2	.28		.32	

1.34, respectively), while baseline year specialty mental health care nearly eliminated it ($O.R.= .13$). R^2 for this model was .32.

Primary Care Clinician's Notation of a Referral to Specialty Mental Health Care

Table 6 shows the final step of hierarchical model for sample members assignable to a personal primary care clinician. Relationship with a personal clinician was the dominant predictor of this outcome. Referrals were similarly elevated among those whose personal clinicians were internal medicine and family practice physicians ($O.R.$ = 3.06 and 2.90, respectively) but strikingly higher among patients of physician assistants ($O.R.$ = 4.90). Second in importance was the interaction term that indicated the sample member had been diagnosed as depressed in the baseline year, but *not* treated with antidepressants. Such persons were nearly three times as likely to be referred by primary care clinicians to specialty mental health care as those who were *neither* diagnosed *nor* treated at baseline ($O.R.$ = 2.84). But being *both* diagnosed *and* treated did not affect referral, nor did being *treated* but *not diagnosed*.

Depressive symptom levels also affected this outcome directly ($O.R.= 1.83$ for moderate symptom levels and 2.27 for high symptom levels) and indirectly (through the higher symptom levels of people [especially women aged 25-54] with a prior mental health care history).

Age and gender interacted to affect the likelihood of referrals. People aged 25-54 were much more likely to be referred than those aged 55 and older. In the 25-54 age group, women's referral rates were double those of men. Among the older sample members, the likelihood of referral was similarly low for men and women ($O.R.$ = 0.53 or less).

Socioeconomic status also significantly predicted referrals. Persons of upper and upper-middle social class were twice as likely to be referred to specialty mental health care as those of middle and lower/working social class. Baseline year referral to specialty mental health care also increased the likelihood of a follow-up year referral ($O.R.= 1.86$). R^2 for this model was 0.19.

Outpatient Visit to Specialty Mental Health Clinician in Follow-up Year

Based on the predictors we considered in this study, our ability to predict this outcome was limited. The first of two hierarchical models we developed included all 7,844 sample members, 208 of whom made at least one follow-up year specialty outpatient mental health care visit. (Results not shown. This model can be viewed at the following website: http://www.kpchr.org/info/present/ dep-care-shye.pdf.) This model did not include Step 8 (relationship to a primary care clinician) because 965 sample members were unassignable. No variables in this model significantly predicted specialty mental health care use and the model's explanatory power was very low (R^2 = .01). The second model included the 6,879

Table 6.   Follow-Up Year Medical Chart Notation of a
Referral to Specialty Mental Health Care by a
Primary Care Clinician $N = 6079$; Number of Events = 95

Variables entered:	Final Step of Model (Step 8)	
Step 1: Social Role Functioning	O.R.	P
High limitation due to emotional problems	1.53	.14
Step 2: Age x Gender		
Men Age 25-54 (Reference)	—	—
Women Age 25-54	2.17	.01
Men Age 55-69	0.52	.11
Women Age 55-69	0.53	.13
Men Age 70+	0.34	.03
Women Age 70+	0.44	.06
Step 3: Level of Depressive Symptoms		
Not depressed on MHI-5 or BSI-8 (reference)	—	—
Depressed on either MHI-5 or BSI-8	1.83	.05
Depressed on both MHI-5 and BSI-8	2.27	.02
Step 4: Mental Health Care History		
Ever told by Dr. you have depression? (Yes)	1.67	.09
Baseline diagnosis of anxiety, sleep/family problems	1.05	.89
Baseline depression Dx x Antidepressant Rx		
Neither Dx nor Rx (reference)	—	—
Both Dx and Rx	1.68	.20
Rx but no Dx	1.55	.45
Dx but no Rx	2.84	.01
Baseline referral to specialty mental health	1.86	.01
Baseline specialty mental health visit	0.73	.45
Step 5: General Health Status		
Chronic Disease Score > 0 (any chronic disease)	0.73	.19
Health now worse than last year? (Yes)	1.49	.15
Step 6: Baseline Health Care Utilization		
High (4+) baseline primary care visits	1.80	.01
Step 7: Sociodemographic Characteristics		
Unmarried	1.01	.96
Socioeconomic status		
Middle/lower/working class (reference)	—	—
Upper middle/upper class	1.97	.007
Step 8: Relation to Personal Primary Care Clinician		
None (reference)	—	—
Internal medicine physician	3.06	.006
Family practice physician	2.90	.01
Physician assistant	4.90	.005
Nurse practitioner	2.61	.11
Summary statistics		
Maximum-rescaled R^2	.19	

sample members whose relationship to a personal primary care clinician could be defined, 188 of whom made at least one specialty mental health visit in the follow-up year. (Results not shown. This model can be viewed at the following website: http://www.kpchr.org/info/present/ depcare-shye.pdf.) Step 8 was included in this model and people whose personal clinician was a family practice physician were marginally less likely to use specialty mental health care than patients of other clinicians or those with no personal clinician ($O.R. = 0.62, p = .07$). Again, the R^2 for the model was very low (.02).

DISCUSSION

We elaborate here on the contribution of our study findings to knowledge about the prevalence and correlates of depressive symptoms, the role of primary care clinicians in depression care (including evidence about the appropriateness of their diagnostic, prescribing, and referral behaviors), the nonmedical factors that affect movement along the pathway to care for depression, and the importance of continuity of care in the treatment of depression.

The Prevalence and Correlates of Depressive Symptoms

Our approach to measuring depression symptom levels does not permit estimation of the prevalence of DSM-defined major depression in our study sample. We found, however, that 14.5 percent of sample members had significant depressive symptoms—that is, they screened positive on at least one of the two depression screeners on which we based our ordinal measure of level of depressive symptoms. This finding is consistent with of other studies. For example, Callahan and colleagues (1994) found that 17.1 percent of a sample of 1,171 primary care patients aged 60 and older had significant depressive symptoms (CES-D score ≥ 16), and Ormel and colleagues (1994), reporting on the WHO Collaborative Study on Psychological Problems in General Health Care, cited the prevalence of ICD-10 depressive disorder to be 15.9 percent, 13.8 percent, and 15.9 percent for the Groningen, Paris, and Rio de Janeiro centers, respectively. As others have (Goldberg and Huxley 1992), we also found a higher prevalence of significant depressive symptom levels among unmarried persons as compared to married, among nonwhites as compared to whites, among unemployed persons (including homemakers) as compared to working or retired people, and among lower socioeconomic class persons as compared to those of higher status.

Our findings are also consistent with other studies (Blazer 1991; Callahan et al. 1994; Unutzer et al. 1997) that indicate that, when significant depressive symptom levels rather than major depression are considered, elderly people are generally similar to younger people. Also, the female excess in the prevalence of significant symptoms in our sample was less striking than that found in some other studies (Culbertson 1997).

The Role of Primary Care Clinicians in Depression Care

Our findings give many indications that clinicians in this HMO are sensitive to and give sustained attention to manifestations of depression in their patients. People who screened positive for depression at the index time point were five times more likely than those who did not to report having "ever been told by a doctor" that they have depression. They were about four times more likely than those who did not screen depressed to have had a baseline year medical chart notation of a depression diagnosis, antidepressant prescription, or referral to specialty mental health care. They were about three times more likely to have had a baseline year chart notation of a mild nondepression psychiatric diagnosis such as anxiety, sleep problems, and family problems, and a serious non-depression psychiatric diagnosis such as alcoholism, personality disorder, or psychosis. That the overwhelming majority of these chart notations were made by primary care clinicians confirms findings from other studies about the dominant role primary care clinicians play in treating depression (Schurman, Kramer, and Mitchell 1984). We further found that primary care clinicians were noting follow-up year depression diagnoses for individuals who had such a diagnosis in the baseline year (who were more likely to screen positive for depression at the index time point than the overall sample), and that they were responsive to their symptoms without regard to age and gender. All these findings appear to reflect a high level of suspicion about the possibility of depression in their patients and sustained attention to it once it is diagnosed. They thus contrast with those of other studies that indicate generally poor recognition of and attention to depression by primary care clinicians (Hays et al. 1989). One explanation for this is that a number of these studies were based on assessment of a single clinician-patient encounter (Wells, Stewart et al. 1989). By contrast, our outcome measures include events over a full year, and may therefore give a better picture of the way primary care clinicians handle the care of depressed patients. Primary care clinicians in this HMO usually have ongoing relationships with their patients and provide care for a wide range of acute and chronic problems. What may be overlooked or less salient at one visit may be noted and dealt with in a subsequent one.

Our findings also shed some light on primary care clinicians' depression care decision-making processes. For example, among people who had no evidence of a recent (baseline year) episode of depression, serious limitations in social role functioning because of emotional problems were an important cue to clinicians' recognition and treatment of depression, independent of level of depressive symptoms.

We examined our findings for evidence about the appropriateness of primary care clinicians' depression care decisions—that is, about whether they were related primarily to patients' levels of depressive symptoms and/or depression-related health and functional limitations, or to other, non-medical, factors. It should be noted that this is a different criterion of appropriateness than that used in studies that have assessed whether clinicians' diagnostic and treatment decisions match research diagnostic criteria (an issue that our study was not designed to address). We found considerable evidence that these primary care clinicians' diagnostic, prescribing, and referral behaviors reflected an appropriate response to their patients' symptoms and history of depression. For example, in recognizing depression and prescribing antidepressants among people who had *no* chart diagnosis at in the baseline year, clinicians appeared to be prompted by their patients' (1) significant functional impairment, (2) current levels of depressive symptoms, and (3) possibility of recurrence of a previous episode of depression. However, among individuals who *had* been diagnosed at baseline, their diagnoses and prescribing appeared to reflect ongoing care for depression. For example, in this subgroup, drug treatment for depression in the baseline year was highly predictive of a primary care clinician's follow-up year depression diagnosis and antidepressant prescription, but level of depressive symptoms was not. This suggests that these clinicians may be appropriately following current guidelines that recommend continued treatment of depressed patients for six months to one year even after remission of symptoms, in order to prevent relapse (Depression Guideline Panel 1993).

Evidence about the appropriateness of primary care clinicians' antidepressant prescribing was mixed. The strong positive effect of high levels of depressive symptoms, past history of medically diagnosed depression, and a baseline year antidepressant prescription all suggested appropriate prescribing. But among persons with *no* baseline-year depression diagnosis, a baseline year chart diagnosis of a mild non-depression psychiatric diagnosis (such as anxiety, sleep problems, or family problems) *also* significantly increased the likelihood of a follow-up year antidepressant prescription. Since these mild psychiatric diagnoses describe symptoms and conditions that frequently precede or accompany depression (Horwath et al. 1994), this may reflect clinicians' efforts to preempt a full-blown episode. Or they may be appropriately recognizing and treating depression but not labeling it as such. Previous research has shown that primary care clinicians frequently avoid giving the diagnostic label to depressed patients, either out of therapeutic considerations or out of concern about possible adverse employment and health care coverage consequences (Goldberg 1982, 1992; Orleans et al. 1985; Rost et al. 1994). A less sanguine interpretation is that primary care clinicians sometimes use antidepressant treatment to make a differential diagnosis. If an antidepressant clears up a confusing presentation of somatic and psychological symptoms, then it was depression; if not, they evaluate further and find another treatment approach. We further found that, among sample members with *no* baseline diagnosis of depression, high baseline year primary care utilization predicted a follow-up year antidepressant prescription. This suggests that primary care clinicians may sometimes use antidepressant prescribing as a way to manage their patient load and deal with troublesome patients.

Primary care clinicians' chart notations of referrals to specialty mental health care were strongly and directly related to patients' depressive symptom levels, and thus, appeared appropriate. An interesting finding regarding referrals was that sample members who in the baseline year were diagnosed as depressed but *not* treated with antidepressants were nearly three times as likely as those who were *neither* diagnosed *nor* treated to be referred to specialty mental health care in the follow-up year; but people who at baseline had *both* a diagnosis *and* a prescription and people who received a prescription *but* no diagnosis were *not* significantly more likely than those with *neither* to be referred. This suggests that primary care clinicians do not refer to specialty mental health care depressed individuals they can treat themselves with antidepressants, but they *do* refer those in whom they recognize depression but can't treat it with drugs—either because they are uncertain about how to treat or because the patients refuse drug treatment or request "talk therapy."

Nonmedical Factors That Affect Care for Depression

Overall we found that patients' depression-related distress and disability and their prior mental health care history were the predominant influences on their clinicians' depression care decisions. But some findings about the role of non-medical factors emerged to qualify this picture of generally appropriate care. In particular, patient gender, age, and/or the interaction between them had some significant direct effects on the outcomes we studied. All the gender effects we observed involved higher rates among women. Among people without a baseline-year depression diagnosis, women aged 25-54 were much more likely than people in any other age/gender strata to be diagnosed as depressed at follow-up. Women in all age groups had higher antidepressant prescription rates than men, even after depressive symptom levels and mental health care history were controlled, and women age 25-54 were more likely than any other age/gender group to be referred to specialty mental health care. These findings are quite consistent with those of other studies (Hohmann 1989; Sprock and Yoder 1997).

We also found consistent age effects that involved differences between people aged 55+ and those aged 25-54: among people with a baseline-year depression diagnosis, those 55+ (regardless of gender) were *more* likely than those aged 25-54 to get a follow-up year diagnosis and a prescription for antidepressants; and in the whole sample, people 55+ were much *less* likely than younger people to be referred to specialty mental health care—also regardless of gender.

Our study data cannot elucidate the mechanisms by which age and gender interact to affect the likelihood of these outcomes independently of people's depression-related characteristics. The explanation is probably related as much to aspects of the clinician-patient interaction as to the patient traits in question or primary care clinicians' depression care knowledge and skills *per se*. For example, it is possible that younger women are more likely than people in other age/gender strata to understand depressive symptoms as medically treatable and demand a response to their symptoms from their primary care clinicians. Thus, clinicians may recognize younger women's depressive symptoms at a lower level of severity than they would among other age/gender groups. Higher rates of depression diagnoses and antidepressant prescribing among elderly people may reflect primary care clinicians' special sensitivity to the risks associated with depression in the elderly—or a reluctance among depressed elderly to use specialty mental health services that makes them more likely than depressed younger people to remain in the primary care system (Goldberg and Huxley 1992; Waxman et al. 1984). Clearly, further study into the mechanisms by which these effects occur is warranted.

Few nonmedical variables other than age and gender had independent effects as predictors of the outcomes we studied. We did find that upper and upper-middle class individuals were significantly more likely than were middle class or lower and working class individuals to receive a referral to specialty mental health care. This finding is consistent with those of other studies that have examined the relationship between socioeconomic status and use of mental health care (Horwitz 1987).

Continuity of Care for Depression

Our study findings make clear the importance for depression care of having an ongoing relationship with a personal primary care clinician. The effect of such a relationship was most striking and consistent when that clinician was a family practice physician: it dramatically increased the likelihood of diagnosis, antidepressant treatment, and referral to specialty mental health care and reduced the chance the sample member would actually use specialty mental health services. The role of relationship to a personal clinician was similar, if less dramatic, among patients of internal medicine physicians. These findings are consistent with those of other studies: family practitioners appear to be more attuned to and better trained than other clinician groups to recognize and treat psychosocial problems and show more therapeutic activism in doing so (Bowman 1990; Glied 1998; Nichols 1998).

In addition, in this HMO setting, specialty mental health care appeared to be a *substitute for* rather than a *complement to* primary care for depression. Specialty mental health care in the baseline year dramatically *reduced* the likelihood of a primary care clinician's depression diagnosis or antidepressant prescription in the follow-up year—irrespective of baseline year depression diagnosis. At least in this HMO setting, there may be a need to enhance the coordination between the two sectors in the care of depressed patients.

Finally, some limitations of our study should be noted. Our findings and the conclusions we base on them come from a single, large, established, vertically integrated, nonprofit group-practice HMO. As such, they probably are generalizable to other similar group- or staff-model HMOs. But a different picture might emerge if the same outcomes were studied in managed care organizations that have less integrated structures or a different set of incentives for patient, clinician, and management behaviors. Current market trends favor less integrated types of managed care such as the independent practice association (IPA) and the preferred provider organization (PPO). Both of these forms of managed care are growing rapidly. Fragmentation of services is a potential problem with these plans and both are less likely than staff and group model HMOs to improve the coordination and continuity of care for mental health problems or other types of disorders. This trend limits the applicability of our findings and our model of care. Also, our conclusions about the general appropriateness of primary care clinicians' depression diagnostic and treatment decisions must be understood in the context of our definition of appropriateness— that is, the extent to which they appeared related to patients' depression-related characteristics rather than to nonmedical factors. Whether these decisions result in optimal outcomes for patients can be determined only by a study that relates specific clinical decisions to patient-specific measures of change in depressive symptom levels and functional status. Another shortcoming of our study is our inability to analyze the effect of *clinician* gender on our study outcomes. Due to strong confounding between gender and clinician type—in particular, the female gender of nearly all nurse practitioners—we were unable to include clinician gender in multivariate models along with "relationship to a personal primary care clinician" (which had the categories: internal medicine or family practice physician, physician assistant, nurse practitioner). Some studies show that women clinicians are more empathic, sensitive and humanistic than men clinicians (Shapiro et al. 1993; Weisman and Teitelbaum 1985), and this may lead to gender differences in clinicians' interest in and skill at diagnosing and treating depression.

CONCLUSIONS

Our study findings have a number of implications for depression care policy, practice, and future research. They further reinforce our awareness of the importance of the primary care system for the care of depression in the community. They suggest that, although primary care clinicians may not be adhering strictly to research diagnostic criteria or mental health specialty depression care guidelines, they may nevertheless be quite sensitive to manifestations of depressions in their patients and provide sustained care that is generally quite appropriate, in that it is related primarily to patients' levels of depressive symptoms, prior mental health care history, and degree of functional impairment, rather than to nonmedical factors.

Our findings also suggest that recognition and treatment of depression in the primary care setting will be enhanced when health care systems link members to a personal primary care physician and create structural arrangements that encourage continuity in patient-clinician relationships. This continuity may be particularly important for elderly people, given the special problems of diagnosing and treating depression in the elderly and the apparent preference—at least in the current cohort of elderly—for primary care treatment for psychosocial problems.

In addition, our findings point to the need for more research into the reasons for and the treatment cost and outcome consequences of differential therapeutic activism in depression care among general internists, family practice physicians, and other primary care clinicians. In particular, attention should be given to identifying the specific aspects of family practice physicians' training and experience that contribute to their greater therapeutic activism in the care of depression and ways should be sought to include these features in the training of other types of primary care clinicians—especially regarding care of the elderly. The effect of clinician gender on depression care diagnostic and treatment activities also deserves further study.

Future research should also continue to explore the reasons for the independent effect of female gender in prompting the diagnosis and treatment of depression. The attitudes and preferences of elderly people regarding depression care also deserve further attention, including how these may be evolving as younger cohorts age.

APPENDIX: DISTRIBUTION OF STUDY VARIABLES

Variable	% Positive (N = 7844)
Outcome variables = events on the pathway to care for depression	
• Use of outpatient primary medical care	77.5
• Medical chart notation by a primary care clinician of:	
• Diagnosis of depression	4.8
• Antidepressant drug prescription	4.8
• Referral to specialty mental health care	1.2
• Outpatient visit to specialty mental health clinician	2.7
Baseline year medical chart notations (by any HMO clinician)	
• Mild psychiatric diagnosis (i.e., anxiety, sleeplessness, family problems)	7.9
• Serious psychiatric diagnosis (i.e., psychosis, suicidality, alcohol addiction)	2.7
• Depression diagnosis	5.3
• Antidepressant prescription	5.7
• Referral to specialty mental health care	1.6
• Depression diagnosis x antidepressant prescription	
• *Neither* depression diagnosis *nor* antidepressant prescription	92.5
• Depression diagnosis *and* antidepressant prescription	3.5
• Antidepressant prescription *but no* depression diagnosis	2.2
• Depression diagnosis *but no* antidepressant prescription	1.8
Baseline year medical care utilization data	
• Office visits to primary care clinicians	
• 0 visits	23.9
• 1-3 visits	47.6
• 4-55 visits	28.5
• Office visits to specialist clinicians (non-mental health)	
• 0 visits	23.9
• 1 - 3 visits	38.8
• 4 + visits	37.3
• Any UC/ER visit in Pre year (yes/no)	27.3
• Any office visit to specialty mental health clinician	2.6
• Non-antidepressant pharmacy dispenses	
• 0 dispenses	19.9
• 1-5 dispenses	30.8
• 6-14 dispenses	26.0
• 15+ dispenses	23.3
• Any antidepressant pharmacy dispense	10.2
• Any inpatient stay	8.9
• Chronic Disease Score	
• 0 (No chronic disease)	47.7
• 1+ (Any chronic disease)	52.3
• 1-2	25.2
• 3-16	27.1

(continued)

Appendix   (Continued)

Variable	% Positive (N = 7844)
Variables from Current Member Survey (measured at index time point)	
• Age/gender	
• Males 25-54	20.3
• Females 25-54	15.3
• Males 55-69	20.0
• Females 55-69	15.6
• Males 70+	12.5
• Females 70+	16.3
• Measures of depressive symptom levels.	
• Not depressed on either MHI-5 or BSI-8 (Low symptom level)	85.5
• Depressed on *either* MHI-5 *or* BSI-8 (Moderate symptom level)	10.2
• Depressed on *both* MHI-5 *and* BSI-8 (High symptom level)	4.3
• Depressed on MHI-5, BSI-8, or both ("Significant depressive symptoms")	14.5
• Marital status	
• Married	75.0
• Not married	25.0
• Ethnicity	
• White	90.6
• Non-white	9.4
• Socioeconomic status (based on self-reported income level and self-reported social class)	
• Lower/working class	28.6
• Middle class	45.0
• Upper middle/upper class	26.4
• Family size	
• 1 person	32.8
• 2 people	45.3
• 3+ people	21.9
• Work status	
• Working	43.7
• Retired	40.0
• Unemployed	16.3
• Level of mental health benefit	
• Standard or less than standard	45.3
• Better than standard	54.7
• History of depression	
• "Has a doctor ever told you that you have depression?" (% Yes)	7.9
• Functional health status (SF-36 scale scores)	
• General health perceptions	
• Low (0-65.79)	35.0
• Medium (67-86.67)	39.7
• High (87-100)	24.3

(*continued*)

Appendix (Continued)

Variable	% Positive (N = 7844)
• Change in health in last year	
Worse than last year (0-25)	12.5
Same as last year (25-50)	68.8
Better than last year (51-100)	18.7
• Degree of body pain	
High pain (0-32)	9.6
Medium (41-84)	47.5
Low pain (100)	42.9
• Vitality (energy/fatigue)	
Low (0-53.33)	35.1
Medium (55-75)	40.6
High (80-100)	24.3
• Level of physical functioning	
• Low (0-72.22)	29.1
Medium (75-94.44)	45.6
High (95-100)	25.2
• Social role limitations due to physical or emotional problems	
High impairment (0-88.89)	41.9
Low impairment (100)	58.1
• Role limitations due to physical health	
Severe limitations (0)	22.4
Some limitations (25-75)	18.9
No limitations (100)	58.7
• Role limitations due to emotional problems	
Severe limitations (0)	12.4
Some limitations (33.33-66.67)	12.7
No limitations (100)	74.8
• Health behaviors	
• Current smoker (% Yes)	16.2
• Alcohol consumption	
• Nondrinkers	34.1
• < 20 drinks/month	51.0
• ≥ 20 drinks/ month	14.9
Relation to personal primary care clinician	
• Has no personal primary care clinician	28.5
• Personal clinician = Internal medicine physician	43.5
• Personal clinician = Family practice physician	20.2
• Personal clinician = Physician assistant	3.6
• Personal clinician = Nurse practitioner	4.2

ACKNOWLEDGMENTS

This research was supported by grant number RO1MH5130502 from the National Institute on Mental Health (NIMH). We wish to thank the following people for their contributions to this project: Jonathan Brown, PhD, Bentson McFarland, MD, PhD, and Richard Johnson, PhD, for their important conceptual and methodological contributions; Greg Nichols, PhD, whose thorough knowledge of the complex data systems on which our study was based and his skill in collecting and organizing the study data sets were crucial for the success of the project; Kim Olson and other research medical information staff for skillfully and carefully abstracting data from study subjects' medical charts; Maggie Vogt for dedicated and skillful analytic support during the critical and complex final phase of our analysis; Jeff Showell for prompt and willing help with data management throughout the project; and Diann Triebwasser and Nance Adams for careful and tireless secretarial help from start to finish.

REFERENCES

Angst, J., and A. Dobler-Mikola. 1984. "Do the Diagnostic Criteria Determine the Sex Ratio in Depression?" *Journal of Affective Disorders* 7: 189-198.

Berwick, D.M., J.M. Murphy et al. 1991. "Performance of a Five-Item Mental Health Screening Test." *Medical Care* 29 (2): 169-176.

Blazer, D.G. 1991. "Clinical Features in Depression in Old Age: A Case for Minor Depression." *Current Opinions in Psychiatry* 4: 596-599.

Boardman, A.P. 1987. "The General Health Questionnaire and the Detection of Emotional Disorder by General Practitioners." *British Journal of Psychiatry* 151: 373-381.

Bowman, M. 1990. "Family Physicians and Internists: Differences in Practice Styles and Proposed Reasons." *Journal of the American Board of Family Practice* 3: 43-49.

Broadhead, W.E., D.G. Blazer et al. 1990. "Depression, Disability Days, and Days Lost from Work in a Prospective Epidemiologic Survey." *Journal of the American Medical Association* 264 (19): 2524-2528.

Brown, J.B., D. Shye et al. 1995. "The Paradox of Guideline Implementation: How AHCPR's Depression Guideline Was Adapted at Kaiser Permanente Northwest Region." *Joint Commission Journal on Quality Improvement* 21 (1): 5-21.

Brown, J.B., D. Shye et al. 2000. "Controlled Trials of Continuous Quality Improvement and Academic Detailing to Implement a Clinical Practice Guideline for Depression." Joint Commission Journal on Quality Improvement 26 (1), in press.

Burnam, M.A., K.B. Wells et al. 1988. "Development of a Brief Screening Instrument for Detecting Depressive Disorders." *Medical Care* 26 (8): 775-789.

Callahan, C.M., S.L. Hui et al. 1994. "Longitudinal Study of Depression and Health Services Use Among Elderly Primary Care Patients." *Journal of the American Geriatrics Society* 42: 833-838.

Cleary, P.D. 1989. "The Need and Demand for Mental Health Services." Pp. 161-184 in *The Future of Mental Health Services Research,* edited by C.A. Taube, D. Mechanic, and A. Hohmann. DHHS Publication No.(ADM)89-1600. Washington, DC: National Institutes of Mental Health.

Cleary, P.D., and B.J. Burns. 1990. "The Identification of Psychiatric Illness by Primary Care Physicians: The Effect of Patient Gender." *Journal of General Internal Medicine* 5: 355-360.

Culbertson, F.M. 1997. "Depression and Gender." *American Psychologist* 52 (1): 25-31.

Depression Guideline Panel (AHCPR). 1993. *Depression in Primary Care: Volume 1. Detection and Diagnosis: Volume 2. Treatment of Major Depression.* Clinical Practice Guideline Number 5. AHCPR Publication No. 93-0550. Rockville, MD: U.S. Department of Health and Human Services, Public Health Service, Agency for Health Care Policy and Research.

Docherty, J.P. 1997. "Barriers to the Diagnosis of Depression in Primary Care." *Journal of Clinical Psychiatry* 58 (Supplement 1): 5-10.

Eisenberg, L. 1992. "Treating Depression and Anxiety in Primary Care: Closing the Gap between Knowledge and Practice." *New England Journal of Medicine* 326: 1080-1084.

Freeborn, D.K., C. R. Pope et al. 1990. "Consistently High Users of Medical Care among the Elderly." *Medical Care* 28 (6): 527-540.

Freeborn, D.K., and C.R. Pope. 1994. *Promise and Performance in Managed Care: The Prepaid Group Practice Model.* Baltimore, MD: The Johns Hopkins University Press. Gallo, J.J. 1997. "Emotions and Medicine. What Do Patients Expect from Their Physicians?" (Editorial) *Journal of General Internal Medicine* 12: 453-454.

Gerber, P.D., J. Barrett et al. 1989. "Recognition of Depression by Internists in Primary Care: A Comparison of Internist and 'Gold Standard' Psychiatric Assessments." *Journal of General Internal Medicine* 4: 7-13.

Gerber, P.D., J.E. Barrett et al. 1992. "The Relationship of Presenting Physical Complaints to Depressive Symptoms in Primary Care Patients." *Journal of General Internal Medicine* 7: 170-173.

Glied, S. 1998. "Too Little Time? The Recognition and Treatment of Mental Health Problems in Primary Care." *Health Services Research* 33 (4): 891-910.

Goldberg, D. 1990. "Reasons for Misdiagnosis." Pp. 139-145 in *Psychological Disorders in General Medical Settings*, edited by N. Sartorius, D. Goldberg, G. De Girolamo et al. Toronto: Hogrefe and Huber.

Goldberg, D. 1992. "A Classification of Psychological Distress for Use in Primary Care Settings." *Social Science and Medicine* 35 (2): 189-193.

Goldberg, D. 1982. "The Concept of a Psychiatric 'Case' in General Practice." *Social Psychiatry* 17: 61-65.

Goldberg, D., and P. Huxley. 1980. *Mental Illness in the Community: The Pathway to Psychiatric Care.* London: Tavistock.

Goldberg, D., and P. Huxley. 1992. *Common Mental Disorders: A Bio-Social Model.* London: Tavistock/Routledge.

Goldberg, H.I., E.H. Wagner et al. 1998. "A Randomized Controlled Trial of CQI Teams and Academic Detailing: Can They Alter Compliance with Guidelines?" *Joint Commission Journal on Quality Improvement* 24 (4): 130-142.

Goldstrom, I.D., B.J. Burns et al. 1987. "Mental Health Services Use by Elderly Adults in a Primary Care Setting." *Journal of Gerontology* 42: 147-153.

Goodman, M.J., D.W. Roblin et al. 1991. "Persistence of Health Care Expense in an Insured Working Population." Pp. in *Advances in Health Economics and Health Services Research, Vol. 12*, edited by M.C. Hornbrook. Greenwich, CT: JAI Press Inc.

Gotlieb, I.H., P.M. Lewinsohn et al. 1994. "Symptoms Versus a Diagnosis of Depression." *Journal of Consulting and Clinical Psychology* 63: 90-100.

Greenberg, P.E., L.E. Stiglin et al. 1993. "The Economic Burden of Depression in 1990." *Journal of Clinical Psychiatry* 54 (11): 405-418.

Greenlick, M.R., D.K. Freeborn et al. 1988. *Health Care Research in an HMO: Two Decades of Discovery.* Baltimore, MD: Johns Hopkins University Press.

Hays, R.D., K.B. Wells et al. 1995. "Functioning and Well-Being Outcomes of Patients with Depression Compared with Chronic General Medical Illness." *Archives of General Psychiatry* 52: 11-19.

Hohmann, A. 1989. "Gender Bias in Psychotropic Drug Prescribing in Primary Care." *Medical Care* 27 (5): 478-490.

62   DIANA SHYE, DONALD K. FREEBORN, and JOHN P. MULLOOLY

Hornbrook, M.C., M.J. Goodman et al. 1996. "Global Risk-Assessment Models: Demographic and Diagnostic Approaches." Paper presented at the Annual Meetings of the Association for Health Services Research, Atlanta, Georgia, June 10.

Horwath, E., J. Johnson et al. 1994. "What Are the Public Health Implications of Subclinical Depressive Symptoms?" *Psychiatric Quarterly* 65 (4): 323-337.

Horwitz, A.V. 1987. "Help-Seeking Processes and Mental Health Services." Pp. 33-45 in *Improving Mental Health Services: What the Social Sciences Can Tell Us,* edited by D. Mechanic. New Directions for Mental Health Services, No. 36. San Francisco: Jossey-Bass.

Johnson, R.E., M.C. Hornbrook et al. 1994. "Replicating the Chronic Disease Score (CDS) from Automated Pharmacy Data." *Journal of Clinical Epidemiology* 47 (10): 1191-1199.

Johnson, R.E., B.H. McFarland. 1994. "Treated Prevalence Rates of Severe Mental Illness among HMO Members." *Hospital and Community Psychiatry* 45 (9): 919-924.

Johnson, J., M.M. Weissman et al. 1992. "Service Utilization and Social Morbidity Associated with Depressive Symptoms in the Community." *Journal of the American Medical Association* 267 (11): 1478-1483.

Katon, W., M. Von Korff et al. 1992. "Adequacy and Duration of Antidepressant Treatment in Primary Care." *Medical Care* 30: 67-76.

Klerman, G.L., M.M. Weissman. 1992. "The Course, Morbidity, and Costs of Depression." *Archives of General Psychiatry* 49: 831-834.

Levin, B.L. 1990. "Mental Health Services within the HMO Group." *HMO Practice* 6 (3): 16-21.

Lin, E., M. Von Korff et al. 1995. "The Role of Primary Care Physicians in Patient Adherence to Antidepressant Therapy." *Medical Care* 33: 67-74.

Marks, J., D. Goldberg et al. 1979. "Determinants of the Ability of General Practitioners to Detect Psychiatric Illness." *Psychological Medicine* 9: 337-353.

McFarland, B.H., R.E. Johnson et al. 1996. "Enrollment Duration, Service Use, and Costs of Care for Severely Mentally Ill HMO Members." *Archives of General Psychiatry* 53: 938-944.

Nagelkerke, N.J.D. 1991. "A Note on a General Definition of the Coefficient of Determination." *Biometrika* 78 (3): 691-692.

National Institute of Mental Health, Depression Awareness, Recognition, and Treatment (D/ART Program). 1987. *Sex Differences in Depressive Disorders: A Review of Recent Research.* Washington, DC: American Psychological Association.

Nease, D.E., Jr., R.J. Volk et al. 1999. "Investigation of a Severity-Based Classification of Mood and Anxiety Symptoms in Primary Care Patients." *Journal of the American Board of Family Practice* 12 (1): 21-31.

Newman, J.P., R.J. Engel et al. 1991. "Age Differences in Depressive Symptom Experiences." *Journal of Gerontology* 46: 224-235.

Nichols, G.A. 1998. "Continuous Detection and Treatment of Depression in a Large HMO." Doctoral dissertation, Portland State University.

Orleans, C.T., L. George et al. 1985. "How Primary Care Physicians Treat Psychiatric Disorders: A National Survey of Family Practitioners." *American Journal of Psychiatry* 142 (1): 52-57.

Ormel, J., M. Von Korff et al. 1994. "Common Mental Disorders and Disability Across Cultures. Results from the WHO Collaborative Study on Psychological Problems in General Health Care." *Journal of the American Medical Association* 272: 1741-1748.

Page, S., S. Benesch. 1993. "Gender and Reporting Differences in Measures of Depression." *Canadian Journal of Behavioral Science* 25: 579-589.

Parchman, M. 1992. "Physicians' Recognition of Depression." *Family Practice Research Journal* 12: 431-438.

Perez-Stable, E.J., J. Miranda et al. 1990. "Depression in Medical Outpatients: Under-Recognition and Misdiagnosis." *Archives of Internal Medicine* 150: 1083-1088.

Pond, C.D., A. Mant et al. 1990. "Recognition of Depression in the Elderly: A Comparison of General Practitioner Opinions and the Geriatric Depression Scale." *Family Practice* 7 (3): 190-194.

Robins, L.N., D.A. Regier. 1991. *Psychiatric Disorders in America: The Epidemiologic Catchment Area Study.* New York: Free Press.

Rost, K., R. Smith et al. 1994. "The Deliberate Misdiagnosis of Major Depression in Primary Care." *Archives of Family Medicine* 3: 333-337.

Schulberg, H.C., W. Katon et al. 1998. "Treating Major Depression in Primary Care Practice: An Update of the Agency for Health Care Policy and Research Practice Guidelines." *Archives of General Psychiatry* 55: 1121-1127.

Schurman, R.A., P.D. Kramer et al. 1984. "The Hidden Mental Health Network: Treatment of Mental Illness by Nonpsychiatrist Physicians." *Archives of General Psychiatry* 42: 89-94.

Schwenk, T.L., J.C. Coyne et al. 1996. "Differences between Detected and Undetected Patients in Primary Care and Depressed Psychiatric Patients." *General and Hospital Psychiatry* 18: 407-415.

Shapiro, J.E., E. McGrath et al. 1983. "Patients, Medical Students' and Physicians Perceptions of Male and Female Physicians." *Perceptual and Motor Skills* 56: 179-186.

Simon, G., J. Ormel et al. 1995. "Health Care of Primary Care Patients with Recognized Depression." *Archives of General Psychiatry* 52 (10): 850-856.

Simon, G., M. Von Korff et al. 1993. "Patterns of Antidepressant Use in Community Practice." *General and Hospital Psychiatry* 15 (6): 399-408.

Sprock, J., C.Y. Yoder. 1997. "Women and Depression: A Update on the Report of the APA Task Force." *Sex Roles* 36 (5/6): 269-303.

Strickland, B. 1988. "Sex-Related Differences in Health and Illness." *Psychology of Women Quarterly* 12: 381-399.

Suchman, E. 1965. "Stages of Illness and Medical Care." *Journal of Health and Social Behavior* 6: 114-128.

Unutzer, J., D.L. Patrick et al. 1997. "Depressive Symptoms and the Cost of Health Services in HMO Patients Aged 65 Years and Older: A 4-Year Prospective Study." *Journal of the American Medical Association* 277 (20): 1618-1623.

Veroff, J., R.A. Kulka et al. 1981. *Mental Health in America: Patterns of Help Seeking from 1957-1976.* New York: Basic Books.

Von Korff, M., E.H. Wagner et al. 1992. "A Chronic Disease Score from Automated Pharmacy Data." *Journal of Clinical Epidemiology* 45 (2): 197-203.

Ware, J.E., and C.D. Sherbourne. 1992. "The MOS 36-Item Short-Form Health Survey (SF-36). I. Conceptual Framework and Item Selection." *Medical Care* 30: 473-483.

Waxman, H.M., E.A. Carner et al. 1984. "Underutilization of Mental Health Professionals by Community Elderly." *The Gerontologist* 24 (1): 23-30.

Weisman, C.S., and M.A. Teitelbaum. 1985. "Physician Gender and the Physician-Patient Relationship: Recent Evidence and Relevant Questions." *Social Science and Medicine* 29: 1119-1127.

Wells, K.B., R.D. Hays et al. 1989. "Detection of Depressive Disorders for Patients Receiving Prepaid or Fee-for-Service Care: Results from the Medical Outcomes Study." *Journal of the American Medical Association* 262 (23): 3298-3302.

Wells, K.B., A. Stewart et al. 1989. "The Functioning and Well-Being of Depressed Patients: Results from the Medical Outcomes Study." *Journal of the American Medical Association* 262 (7): 914-919.

SOCIAL RELATIONSHIPS AND PSYCHIATRIC HELP-SEEKING

Anthony C. Kouzis, Daniel E. Ford, and
William W. Eaton

ABSTRACT

The effects of distress and social networks on psychiatric help seeking were examined in an adult sample from a community survey of 3,481 adults in Baltimore,
Maryland. Data were derived from the Johns Hopkins University site of the NIMH
Epidemiologic Catchment Area program. Statistical adjustment for the independent
effects of social (age, education, marital status, race, household composition, and
sex), economic (employment, income, and insurance), and physical health factors
were controlled for in estimating the relative odds of mental health service utilization. Subjects who were young, without full-time employment, or who reported one
or more chronic medical problems were more likely to utilize mental health professionals. Married persons and the aging were less likely to seek psychiatric treatment.
Social support and psychological distress interact to affect the use of mental health
care. Persons with weak family ties were five times more likely to seek professional
help than those with strong family ties, while persons with confiding social support
were over four times as likely to use mental health services as those lacking
confiding relationships. Interventions and other treatment efforts to encourage use
of mental health services are recommended.

Research in Community and Mental Health, Volume 11, pages 65-84.
Copyright © 2000 by JAI Press Inc.
All rights of reproduction in any form reserved.
ISBN: 0-7623-0671-8

Although most persons experiencing emotional distress do not receive treatment (Shapiro et al. 1984; Leaf et al. 1985; Howard et al. 1996), the use of outpatient mental health services has grown rapidly since the 1970s (Wells et al. 1982; Olfson and Pincus 1994; Kessler et al. 1999). While a variety of factors affect the decision to seek help for emotional problems (Regier, Goldberg, and Taube 1978; Leaf et al. 1988; Robins and Regier 1991; Wu, Kouzis, and Leaf 1999), their relative importance remains unclear. In particular, the association between social relationships, psychosocial stress and mental health services use is not well understood, despite many analyses on help seeking behavior and social support.

SOCIAL NETWORKS AND SOCIAL STRESS

Numerous investigations have examined the effects of social factors on distress, physical and psychiatric morbidity and mortality (Durkheim 1951; Berkman and Syme 1979; House, Robbins, and Metzner 1982; Cohen and Wills 1985; Mirowsky and Ross 1989). Social support is linked to stress and illness (Gore 1978; House 1981; Cohen and McKay 1984) by either intervening to prevent a stress appraisal response or by reducing or eliminating the stress reaction. Cassell (1974), Cobb (1976), and Kaplan, Cassell, and Gore (1977) have integrated social support into the stressor illness model. Most studies of social support and health have focused *primarily* on two types of effects: *main effects* (Andrews et al. 1978; Lin et al. 1979; Henderson et al. 1980; Williams, Ware, and Donald 1981; Sherbourne 1988; Roberts, Dunkle, and Haug 1994) in which the positive association between social support and well-being is attributed to an overall beneficial effect of social support; and second, a *buffering effect* (Nuckolls, Cassell, and Kaplan 1972; Brown, Bhrolchain, and Harris 1975; Eaton 1978; Turner 1981; Thoits 1982; House and Kahn 1985; Lin, Woelfel, and Light 1985) in which support protects persons from the potentially adverse effects of stressors (Cohen and Wills 1985; Pearlin and Schooler 1978; Finney et al. 1984).

Family members are known sources of social support (Rossi and Rossi 1990), often at personal and economic costs to other members of the family (Fisher, Benson, and Tessler 1990). Kin helping may be a familiar coping mechanism in times of pernicious life events; mothers are most likely to serve as central caregivers for the severely mentally ill (Horwitz and Reinhard 1995). In Ghana, Fosu (1995) found that family size influenced women's orientation toward hospitalization for mental illness; large social networks were seen as easing access to treatment or providing information regarding treatment. Berkman and Syme (1979) found family and friendship were positively associated with health behaviors and preventive health services.

Friends are singularly important to one's sense of well-being (House and Kahn 1985; Seeman and Berkman 1988). Sherbourne (1988) found high social support (number of contacts) was positively related to the likelihood of services use.

Langner and Michael (1963) assert that the crucial distinction is between having no friends and having one or more, while Eaton (1978) and Kessler and McLeod (1985) suggest that a perceived intimacy or closeness with a confidante is particularly crucial (cf. House and Kahn 1985). More recently, Thoits (1992) and Laireiter and Baumann (1992) have shown how perceived social support usually produces main and buffering effects while network tie measures show mostly main effects. In an interesting review, Lin and Peek (1999) argue that conflicting findings may result from a failure to include dynamics and process of support and conditions of life course and role relationships in our analysis.

SOCIAL RESOURCES

Vaux (1988) and Turner and Marino (1994) focused on the social distribution of social support. Evidence exists for continued *sex* (Gove and Tudor 1973; Wells, Manning, and Benjamin 1986; Leaf and Bruce 1987; Verhaak 1995; and Fortney et al. 1998), *race* (Hough et al. 1978; Leaf et al. 1985; Sussman, Robins, and Earls 1987; Padgett et al. 1994; Gallo et al. 1995; Neighbors and Jackson 1996) and *age* (Shapiro et al. 1984; Horgan 1986; and Mechanic, Angel, and Davies 1992) differences in use of mental health services. Living *alone* is associated with a greater likelihood of using the mental health sector (Link and Dohrenwend 1980; Fox 1984; Leaf et al. 1988). Unmarried persons were also more likely to seek help for emotional distress (Fox 1984; Leaf and Bruce 1987; Golding and Wells 1990; Ross 1995; Waite 1995). *Structural* factors such as income, health insurance (public or private), and availability of health care facilities are known to be important. Employed persons have been less likely to use mental health services than those who were unemployed (Olfson and Klerman 1992; Verhaak 1995). Those with higher income (Dutton 1978; Wells, Manning, and Benjamin 1986; Greenley, Mechanic, and Cleary 1987; Katz et al. 1997) or more insurance (Catalano, Rook, and Dooley 1986; Sherbourne 1988; Landerman et al. 1994; Bongers et al. 1997; Fortney, Rost, and Zhang 1998) were more likely to receive mental health services. *Education* has been found to be important. Fijhuis, Peters, and Foets (1990) reported a greater likelihood of using specialty sector with increased education. Mechanic and colleagues (1992) revealed education to be the sole sociodemographic correlate for either diagnosis or a variety of mental health services (medications or psychotherapy). *Physical health* factors have been the most powerful predictors of health care use and include such variables as symptoms, disability, or chronic diseases (Wolinsky 1976; Leaf et al. 1988; and Wells, Burnam, and Camp 1995). *Distress* increases the probability of using medical care and "scales of psychological distress and reporting of common physical symptoms are substantially correlated (and) have many similar predictors" (Mechanic 1979). *Mental health* need has been the most important predictor of mental health services utilization (Greenley and Mechanic 1976; Leaf et al. 1985; Portes, Keyle, and Eaton 1992).

There are few studies of mental health services use and *social relationships*; findings of those that do exist are often unclear (House, Umberson, and Landis 1988; Lin and Peek 1999). Horwitz (1977) found less reciditivity among community mental health patients favored with support from family members. Analysis of the Los Angeles ECA by Golding and Wells (1990) showed Mexican-American and nonHispanic whites with low support from a spouse or other relatives reported a greater likelihood of seeking formal care. Similarly, Catalano, Rook, and Dooley (1986) asserted persons with high support from friends and coworkers were more likely to use services during times of economic insecurity. However, Sherbourne's (1988) analysis of the Rand Health Insurance Experiment demonstrated "neither social contacts nor social resources buffer the impact of life stress events on use of services." Confiding social support, social participation, and social contact were not significantly associated with use of mental health services in a community survey by Greenley, Mechanic, and Cleary (1987).

There are three limitations of existing studies. First, most research on the effects of social stress and social support has focused on health or mental health status as an outcome variable. Two, information concerning which aspects of social networks and their supportive functions (i.e., social support) are related to utilization is sparse. Within the stress literature, social network measures have not received as much focus as perceived social support measures (Thoits 1992; Glass et al. 1997; Pescosolido and Boyer 1999). Three, analyses of stress *and* social support to explain services use have examined *general* health services use among adult (Pilisuk, Boylan, and Acredolo 1987; Broadhead et al. 1989; Counte and Glandon 1991) or pediatric (Horwitz, Morgenstern, and Berkman 1985) patient samples. Those examining community samples have focused on the aging (Krause 1988; Wolinsky and Johnson 1991) or minorities (Pescosolido et al. 1998). There are little data examining the influence of distress, social relationships, and social support on mental health services utilization in the general population.

HYPOTHESES

Herein we examine the stress-buffering effects of social support on help seeking for mental health. We assess the possibility that social support interacts with psychological distress in affecting the use of mental health services. The presence of high distress is hypothesized to promote help seeking while low distress is expected to decrease use of services. For support the prediction is more complicated: while shared confidences may help reduce formal help seeking, supportive social relationships may guide the individual to use health services. In analyzing the interaction, the main effects of social support and distress are considered. We predict that social support will not, in and of itself, lead to higher use of services.

The hypothesized direction of the interaction is that high distress, when combined with the absence of social support resources will lead to a greater likelihood of using mental health services. In this situation, the individual will have a higher than usual tendency to seek help.

These relationships are analyzed while controlling for the potential differences from significant social (age, sex, education, household size, marital status, and race), economic (income, employment), and health (insurance, physical health) resources. We synthesize social support and social networks to show how social resources influence help seeking behavior. Following Pescosolido (1992) events such as help seeking are viewed as *network* rather than individual phenomena. Services use is the "pattern and pathways of practices and people consulted in an illness episode"; our concern is with the dynamics of how people seek help. We seek to understand the social factors in the process of coping with mental illness (cf. Pearlin 1999).

METHODS

ECA Design

These analyses are conducted with data from the NIMH Epidemiologic Catchment Area (ECA) Program (Eaton and Kessler 1985; Robins and Regier 1991). The ECA Program is a series of epidemiologic surveys conducted by university-based researchers in five community mental health center catchment area populations. At each site interviews were conducted with probability samples of about 3,000 individuals living in household populations (Eaton et al. 1984). This paper reports analyses of data from the Johns Hopkins University site of the ECA Program. The Baltimore ECA site collected data on social support and chronic medical conditions which allow a richer analysis then is possible in the five ECA-site data base. The sample of 3,481 individuals at this site was drawn from the population of East Baltimore, comprising some 175,000 adults in 1981. The response rate was 78 percent (Von Korff et al. 1985).

Variables

One of the original purposes of the ECA Program was to describe patterns of *health care utilization* in the population, with a special emphasis on the interrelationship between mental health and general health facilities (Eaton et al. 1981; Regier, Goldberg, and Taube 1978). A special section of the questionnaire was designed specifically for this purpose (Shapiro et al. 1985). The questions resemble those used in the Health Interview Surveys (HIS) and the Medical Care Utilization and Expenditure Surveys (MCUES) of the National Center for Health Statistics, in which the respondent is asked to recall and report use of services.

The ECA questions differ from the HIS and MCUES in extending the recall period for use of services to six months, and in having much more detail in use of mental health services. Earlier reports have presented descriptive data on the use of health and mental health facilities in East Baltimore (Shapiro et al. 1984; Shapiro et al. 1986). Variables for health services use in this presentation are identical to those used in earlier reports.

Several sets of variables included in the Eastern Baltimore site of the ECA Program were not included in other ECA sites. The Baltimore site included the General Health Questionnaire (GHQ) in order to measure *distress* that might not be connected to a specific mental disorder. The GHQ was designed for study of the relationship of psychiatric problems to general health care utilization, and has good psychometric properties (Goldberg 1972). The GHQ questions are framed with respect to the respondent's "normal" performance over the last several weeks, for example, item___: "Over the past several weeks have you been more upset than usual?" The standard threshold of four or more was used to define those in the sample considered to be under "high distress," while those scoring less than four were considered to be at "low distress." Over 16 percent (*n* = 551) of our sample received a high distress score.

Three questions, which ask about contact with family, friends, and existence of a confidante, measure social networks and social support. The *confidante* question for social support asks, "If you had a very personal and serious problem, are there any people with whom you could discuss it?" Two questions ask about personal network ties. Regarding contacts with *family*, respondents were asked: "How many family members and relatives, who do not live with you, do you usually keep in touch with by telephone or by visiting? Include those you have kept in touch with during the past six months." We assumed family members were present if respondents stated they were living in a household. The measure for contacts with *friends* states: "This question has to do with your friends. These can include neighbors, people you work with, and someone else you consider a friend. How many friends like these do you keep in touch with by telephone or visits?" The responses to the support variable and social network variables were dichotomized by recoding responses as "zero" or "one versus greater than one."

The *sociodemographic* variables included: age, sex, race, education, household size, and marital status. *Economic* factors included employment status, income level in dollars, and insurance status. Full-time employment status was ascertained by asking respondents what they were doing most of last week (as in the Labor Force surveys). All responses other than full time employment were coded "not employed." Income included salaries, wages, social security, welfare, and any other income. Insurance refers to private or publicly assisted forms of payment for health care services. Private forms of insurance include health insurance plans paying part or all doctor bills, health maintenance organizations (HMOs) or prepaid group practice plans. Publicly assisted medical assistance or other public programs such as welfare or public assistance forms of health care as well as Medicare from Social Security were also considered.

The presence of any *chronic health conditions* was drawn from the Health Interview Survey. Respondents were queried as to whether they were currently experiencing specific physical health conditions (asthma, high sugar or diabetes, heart trouble, high blood pressure, arthritis or rheumatism, trouble breathing, a stroke, and/or cancer). Those answering "no" to all of these conditions were coded "0", while a yes response to any of the above was a "1." Psychological distress was the second health measure.

Use of *mental health care services*, our dependent variable, was operationalized to include any visit to a general medical doctor or specialty mental health professional for a mental health reason in the previous six months. It includes mental health visits to general medical and human service sector in which the major purpose of the visit was for problems of nerves, emotions, and/or drugs or alcohol and/or specialty care settings, but does not include visits to social workers and counselors in other settings. Over 7 percent (*n* = 252) reported such a visit (Shapiro et al. 1985).

Sample

The sample design for the ECA surveys included several stages, with clustering of units for selection at one or more stages (Eaton et al. 1984). The purpose of this analysis is analytic, not descriptive—in effect, considering variation within this sample of individuals. Therefore, the analyses are not weighted by sample selection probabilities. In estimation of variances, the sample is treated as a simple random sample, instead of a multistage clustered sample. The result is that estimates of variance are likely to have a low bias with the design effect, across a variety of variables between 1 and 1.3. The reader should be aware of this in considering the statistical significance of results (Kessler et al. 1985). The analyses below use multiple logistic regression procedures to model the odds that an individual will report one or more visits to a health care facility for mental health services during the six months prior to the interview.

RESULTS

Frequency distributions for predisposing and enabling factors are considered (Table 1). Sixty-three percent (*n* = 2,159) of our respondents are women. About one third (*n* = 1,182) are African American. Forty percent (*n* = 1,411) are married; 22 percent (*n* = 777) live alone. Most have insurance/HMO and/or are eligible for Medicaid or Medicare (74 percent, *n* = 2,564). Forty-four percent (*n* = 1,539) earn low- or moderate-income while 46 percent (*n* = 1,463) are currently employed full-time.

Table 1. Description of Baltimore ECA Sample

	Number	*Percent*
Sex		
Male	1,322	38
Female	2,159	63
Age		
18-29	925	27
30-44	791	24
45-64	842	24
65+	923	27
Education		
Eighth grade or less	990	28
Ninth to Eleventh grade	896	26
High School	948	27
Beyond High School	646	19
Race		
African American	1,182	34
White and Other	2,299	66
Marital Status		
Married	1,411	40
Not Married	2,069	60
Household Size		
One	777	22
Two	1,163	32
Three	635	19
Four or more	906	26
Insurance		
Insurance	2,564	74
No Insurance	917	26
Income		
Under $4,999	621	18
$5,000-12,499	918	26
$12,500-19,999	592	20
$20,000 +	862	29
Employment		
Full-time	1,463	46
Not Full-time	1,904	54
Medical Illness		
One or more	1,586	45
None	1,895	54

Note: The total sample consists of 3,481 individuals. Data were required to be complete for age, sex, and race. For other variables the total does not always equal 3,481 owing to missing data.

Table 2. Predictors of Mental Health Visits

Factors	% Respondents With Mental Health Visits	Odds+ of Visiting	Adj Odds Ratios++ of Mental Health Visits
Sex			
Male	6	1.00	1.00
Female	8	1.40	1.11 (0.83, 1.49)
Age			
18-29	8	1.00	1.00
30-44	9	1.43	**1.87 (1.32, 2.66)**
45-64	9	1.39	1.40 (0.98, 1.98)
65+	5	0.31	**0.50 (0.36, 0.71)**
Marital Status			
Married	5	1.00	1.00
Not Married	9	0.56	**0.55 (0.40, 0.75)**
Household Size			
One	8	1.20	1.38 (0.90, 2.10)
Two	7	0.90	**1.52 (1.04, 2.22)**
Three	9	1.38	**1.60 (1.07, 2.38)**
Four or more	6	1.00	1.00
Employment			
Full-time	6	1.00	1.00
Not full-time	9	1.53	**1.86 (1.37, 2.51)**
Medical Illness			
One or more	9	1.00	1.00
None	6	1.60	**1.52 (1.14, 2.04)**

Notes: + *n* varies by variable as in Table 1.
++ Adjusted for all variables in the table. The number of observations is *n* = 3,366 owing to missing data.
Bold-faced odds ratios are those whose 95% Confidence Intervals do not include 1.00.

The effect of significant sociodemographic factors to mental health visits is presented in Table 2. The range in visitation rates is 5 to 9 percent (column 1). Aging and married persons show the lowest rates while middle aged, unmarried, unemployed and chronically ill persons report the highest (column 2). Respondents with a chronic medical problem or without full-time employment have the highest crude odds of visiting mental health facilities. These are not statistically adjusted for other covariates, however.

Confidence intervals were obtained from a logistic regression model with the above variables to ascertain the net effect of each variable on the use of mental health services. Statistically significant covariates were being young, being unemployed, and having a chronic medical problem. Living in smaller households was significant for three (OR = 1.60, 95% CI = 1.07, 2.38) and for two person (OR = 1.52, 95% CI = 1.04, 2.22) households. The reduced relative odds of mental health services use for the aging group (OR = 0.50, 95% CI = 0.36, 0.71), and married persons (OR = 0.55, 95% CI = 0.40, 0.75) were also significant.

Table 3. Multivariate Logistic Regression of Social Network Predictors of Mental Health Visits

| | | Adjusted Odds Ratios | | |
| | | Distress and Social Network/Support Interactions Models+ | | |
	1 Main Effects	2 Confid.	3 Friends	4 Family
Distress				
Low	1.00	—	—	—
High	4.35***	—	—	—
Confidante				
No Confidante	1.00	—	1.00	1.00
Confidante	0.89	—	0.82	0.88
Contact with Family				
No Contact	1.00	1.00	1.00	—
Some Contact	1.93	1.91	1.01	—
Contact with Friends				
No Friends	1.00	1.00	—	1.00
Some Contact	1.10	1.07	—	1.08
Distress by Confidante				
Low Support, Low Distress	—	1.00	—	—
High Support, Low Distress	—	0.87	—	—
High Support, High Distress	—	4.04***	—	—
Low Support, High Distress	—	2.70*	—	—
Distress Support by Friends' Contact				
No Contact, Low Distress	—	—	1.00	—
Contact, Low Distress	—	—	0.59	—
Contact, High Distress	—	—	2.76**	—
No Contact, High Distress	—	—	1.66	—
Distress Support by Family Contact				
No Contact, Low Distress	—	—	—	1.00
Contact, Low Distress	—	—	—	0.70
Contact, High Distress	—	—	—	1.77
No Contact, High Distress	—	—	—	5.12***
N =	2611	2611	2611	2611

Notes: + Adjusted for age, sex, marital status, household size, employment, and medical illness
* $p < .05$
** $p < .01$
*** $p < .001$

The relative odds of mental health services use by different levels of distress, the effects of social support provided by confidantes and by social contact with family members or with friends are examined (Table 3). The interaction effects of social network ties and distress are also considered. To do so, we created three indicator variables, which together represent the information in a two by two cross-classification of social support/network (low versus high) and distress (low versus high). The reference category is the group with low distress and low social support/network, which should have a low rate of psychiatric care. The other three categories are high social support/network and low distress, high social support/ network and high distress, and low social support/network and high distress. These models for confidante, friends, and family appear in columns 2, 3, and 4, respectively. For purposes of comparison and compatibility of results we restricted our sample to the number in the regression equation of Model 1 ($n = 2,611$); 840 persons were missing on at least one variable.

The main effects of distress and social support or social network on mental health utilization are also presented (Model 1). Model 2 includes the interaction terms for social support provided by a confidante. In this model, distress and confidante are removed as main effects from our regression model because they are redundant with the interaction terms. In Models 3 and 4 the main effects of network ties with friends and network ties to family respectively are removed for the interaction terms created by their product. The coefficients are adjusted for age, sex, education, marital status, household size, employment and chronic health problems; these were selected on the basis of their statistical significance in Table 2.

The equation including main effects of distress and social support on mental health care (Model 1) illustrates quite strongly the association of distress with mental health visits. The group reporting high distress on the GHQ had a 4.4 higher odds (95% CI = 3.16, 5.99) of having a mental health visit, even after controlling for our sociodemographic variables. Of the three types of social relationships examined, only contact with family appears strong with increased odds of visiting a mental health professional twice those not seeking care. However, this did not reach conventional levels of significance. We hypothesized that social support alone would not influence health utilization.

Presented in Model 2 is the test of the hypothesis of the buffering effects of social support. Specifically, we examined the interaction of distress and having a confidante. The odds of using psychiatric services by persons with confidantes and high distress is four times greater (O.R. = 4.04, 95% CI = 2.04, 7.99) than for the reference group. This is as expected in our original hypothesis. The odds of seeing a mental health professional for those lacking a confidante and experiencing high distress are almost three times as great (O.R. = 2.70, 95% CI = 1.06, 6.87) as the comparison group with low distress and no social support. The greater relative odds of seeking help for those with high levels of distress irrespective of the presence of emotional support may stem from either of the two models. The first, the "psychosocial stressor effect" may promote treatment for those without support. The second, the "treatment facilitating effect" may guide those with close emotional support from friends to seek professional help.

Examining the measures for social network ties with friends or with family allows a closer look at the social source of support. We also examine the interaction of high symptoms of distress with social network ties with friends (Model 3). For those with contact with friends and high distress the relative odds of using health services are twice as high (O.R. = 2.76, 95% CI = 1.31, 5.83) as the reference group. We also examine the interaction of family contact with distress (Model 4). A five times (O.R. = 5.12, 95% CI = 3.56, 7.36) greater relative odds of seeking professional services by those with high distress and no contact with family members was found. Thus, the results from both models 3 and 4 were consistent with our expected hypotheses.

CONCLUSION

The high levels of symptomatology for psychiatric disorders and distress evidenced by the large epidemiologic surveys of the U.S. population in the 1980s and 1990s (Robins and Regier 1991; Kessler et al. 1994; Eaton et al. 1997) and concomitant increase in utilization of specialty mental health services or use of psychiatric medications have not reduced the disparity (Regier et al. 1993; Narrow et al. 1993) between need and levels of mental health services use. This suggests that diagnosis is not a perfect indicator of need, or that need is not a sufficient determinant of the use of mental health services in society (Regier et al. 1998) despite the preponderance of research data asserting this to be the case (Mechanic 1979; Anderson 1995).

Reflecting on his model of health behavior, Anderson (1995) has recently asserted that social networks "rightly fit" into the analysis of health services use. In prior research, Kouzis and Eaton (1998) examined the influence of distress, absence of social support and utilization for services in the general medical sector. In this analysis we explore these relationships for help seeking in the psychiatric services sector. Respondents who were not working, unmarried, or young were more likely to utilize mental health services. Those with chronic physical health problems were also more likely to seek help from mental health professionals. In contrast, persons who were older or married were less likely to seek psychiatric help. These factors exerted independent effects on the likelihood of contact with a mental health professional. Our findings support the inclusion of distress and social relationships in explanations of health services utilization. Models of buffering effects (Eaton 1978; Cohen and Wills 1985) using measures of perceived availability of emotional support in times of distress with mental health services as the dependent variable are also supported.

Attempts at coping with distress are created in negotiation with others and constrained by social structure. The most important finding appears for those with weak social network ties to relatives who were found to be five times more likely to utilize mental health services during times of high distress than people with strong ties with family members. The *absence* of kin may be a source of distress ("stressor-facilitating effect") or family members may simply discourage treatment for fear of stigma (Goffman 1961; Rosenfield 1997) or burden of care (Fisher et al. 1990; Lefley 1996). Persons *with* kin ties are often assisted informally outside of the formal mental health sector. Family members are known to be less likely to have stigmatizing attitudes toward people with mental illness than more distant members of their social networks (Link 1987; Horwitz and Reinhard 1995). While family ties are typically more permanent and reliable (Rossi and Rossi 1990), friends, in contrast, may steer those experiencing distress to seek care by identifying problems and providing referrals to professionals; the "treatment facilitating effect" of social support (cf. Kadushin 1969). While our data preclude further explanation, the greater objectivity of friends or the normative requirements of the workplace may be involved. Persons experiencing troubling events may be encouraged to seek help in times of need, turning to professionals when assistance is *not* found within the social network (Kasl, Gore, and Cobb 1975). Recent evidence on the social networks of African American women (Mays, Caldwell, and Jackson 1996) and working class residents of Puerto Rico (Pescosolido et al. 1998) suggests large supportive social networks among minority groups provide informal support. This may result in a decreased likelihood of resorting to providers in the formal mental health sector. Alternatively, Kadushin (1969) and Sherbourne (1988) found high social support from friends and family was positively related to the likelihood of services use.

A better understanding of the social relations of help seeking and the process of coping with distress may assist in developing better intervention strategies to reduce barriers to the use of mental health services (cf. Pescosolido and Kronenfeld 1995). Intervention efforts could focus on reducing the social isolation of patients and the fostering of closer emotional ties (cf. Rook 1984), particularly with family members (Keitner et al. 1995). Treatment efforts should incorporate social as well as clinical attributes of those seeking professional care (Kiesler 1985). For example, evidence from clinical trials in Britain by Tyrer and associates (1994) has shown how efforts to increase well being by fostering large social networks have been more successful for patients with some disorders but not others. Developing interventions based on the data from this study are premature. Patients with low social support may be appropriate targets for programs trying to match mental health services to those with the greatest need. Moreover building social support may reduce utilization for those with their first episode, or more importantly, recurrent episodes of psychiatric distress.

Our emphasis on the interaction between social networks, distress, and services utilization has not included an investigation into types of psychopathology, when medications such as antidepressants substitute for psychotherapy, or whether one is seeking help for psychiatric distress from a general medical doctor, psychologist, or a psychiatrist. A shortcoming in our analysis is the limitation of cross-sectional analysis and the relative paucity of our items measuring social networks and social support. Panel studies with more detailed measures of social ties would allow an examination of the expanded social model in greater detail. In presenting our findings, we have been cautious because the sampling design was complex but estimates of variance did not take account the clustered nature of the sample, therefore, variances are understated (but probably not by more than 25 percent).

In the past decade, research has focused more on improving outcomes for individuals using treatment services, rather than understanding and improving the process by which individuals access mental health services. Our analysis documents the impact of social support on seeking care for psychiatric distress. More interesting is our data suggest that lack of contact with friends may not lead to as large an increase in use of mental health services as lack of contact with kin. Of course, data on outcomes would provide a more complete picture of how social support should be incorporated into future interventions. These findings do suggest that as families become smaller and more adults live alone, use of mental health services should increase (cf. Kramer 1983). The increasing life span for persons, particularly among the seriously mentally ill who are often alone, will leave care to siblings as parents become disabled or die (Horwitz 1992). Moreover, as care for mental health continues to be centered in the community (Morrissey and Goldman 1986) rather than the asylum (Brown 1985; Sutton 1991), as managed care leads to reduced types and frequency of treatment (Wells et al. 1986; Sturm et al. 1995), and as the proportion of aging or disabled (Kramer 1983) in the population increases, burdens of support and care for persons with psychiatric distress will increase for family and confidantes.

NOTES

1.   The data were reanalyzed with weighted data to compensate for the lack of representativeness of the sampling scheme. Trivial differences were found utilizing a weighting scheme: aging 0.17 (CI = 0.06, 0.52), unemployed 2.07 (CI = 1.19, 3.6), and high distress 4.59 (CI = 2.71, 7.78) had the strongest association with mental health services use. In the base model, chronic physical condition reaches significance for the first time with an odds 1.77 (CI = 1.10, 2.85) times that of the reference group. In the interactive models, frequent contact with friends and high distress loses significance and reduces odds of visiting.

2.   We replicated our main effects model for those with multiple mental health visits. For this group (n = 81), being young (O.R. = 3.05, CI = 1.65,5.65), old (O.R. = 0.32, CI = 0.16, 0.64), living in poverty level household (O.R. = 3.12, CI = 1.43, 6.81), and not working full-time (O.R. = 2.03, CI = 1.12, 3.69) were significant predictors of multiple visits.

ACKNOWLEDGMENTS

The National Institute of Mental Health (grant number MH47447) supported this research. Morton Kramer provided valuable insights and encouragement. The authors thank Peggy Thoits for her criticisms of an earlier version of the paper.

REFERENCES

Andersen, R. M. 1995. "Revisiting the Behavioral Model and Access to Medical Care: Does It Matter?" *Journal of Health and Social Behavior* 36: 1-10.

Andrews, G., C. Tennant, D. Hewson et al. 1978. "Life Event Stress, Social Support, Coping Style, and Risk of Psychological Impairment." *Journal of Nervous and Mental Disease* 166: 307- 316.

Berkman, L.F., and S. L. Syme. 1979. "Social Networks, Host Resistance, and Mortality: A Nine-Year Follow-up Study of Alameda County Residents." *American Journal of Epidemiology* 109: 186-204.

Bongers, I. M. B., J. B. W. Van Der Meer et al. 1997. "Socio-Economic Differences in General Practitioner and Outpatient Specialist Care in the Netherlands: A Matter of Health Insurance?" *Social Science and Medicine* 44: 1161-1168.

Broadhead, W. E., S. H. Gehlbach, F. V. DeGruy et al. 1989. "Functional versus Structural Social Support and Health Care Utilization in a Family Medicine Outpatient Practice." *Medical Care* 27: 221-233.

Brown, G. W., M. N. Bhrolchain, and T. Harris. 1975. "Social Class and Psychiatric Disturbance among Women in an Urban Population." *Sociology* 9: 225-254.

Brown, P. 1985. *The Transfer of Care: Psychiatric Deinstitutionalization and Its Aftermath.* London: Routledge & Kegan Paul.

Cassell, J. C. 1974. "An Epidemiological Perspective on Psychosocial Factors in Disease Etiology." *American Journal of Public Health* 64: 1040-1043.

Catalano, R., K. Rook, and D. Dooley. 1986. "Labor Markets and Help Seeking: A Test of the Employment Hypothesis." *Journal of Health and Social Behavior* 27: 277-287.

Cobb, S. 1976. "Social Support as a Moderator of Life Stress." *Psychosomatic Medicine* 38: 300-314.

Cohen, S., and G. McKay. 1984. "Social Support, Stress and the Buffering Hypothesis: A Theoretical Analysis." Pp. 253-267 in *Handbook of Psychology and Health*, vol. 4, edited by A. Baum et al. Hillsdale, NJ: Erlbaum.

Cohen, S., and T. A. Wills. 1985. "Stress, Social Support, and the Buffering Hypothesis." *Psychological Bulletin* 98: 310-357.

Counte, M., and G. L. Glandon. 1991. "A Panel Study of Life Stress, Social Support, and the Health Services Utilization of Older Persons." *Medical Care* 29: 348-361.

Durkheim E. 1951. *Suicide.* New York: Free Press.

Dutton, D. B. 1978. "Explaining the Low Use of Health Services by the Poor: Costs, Attitudes, or Delivery Systems." *American Sociological Review* 43: 348-368.

Eaton, W. W. 1978. "Life Events, Social Supports, and Psychiatric Symptoms: A Re-analysis of the New Haven Data." *Journal of Health and Social Behavior* 19: 205-229.

Eaton, W. W., J. Anthony, J. Gallo et al. 1997. "Natural History of DIS/DSM Major Depression: The Baltimore ECA Followup." *Archives of General Psychiatry* 54: 993- 999.

Eaton, W. W., C. E. Holzer, Michael Von Korff et al. 1984. "The Design of the Epidemiologic Catchment Area Surveys." *Archives of General Psychiatry* 41: 942-948.

Eaton, W. W. and L. G. Kessler (Eds.). 1985. *Epidemiologic Field Methods in Psychiatry: The NIMH Epidemiologic Catchment Area Program.* Orlando, FL: Academic Press.

Eaton, W. W., D. Regier, B. Z. Locke et al. 1981."The Epidemiologic Catchment Area Program of the National Institute of Mental Health." *Public Health Reports* 96: 319-325.

Fijhuis, M., L. Peters, and M. Foets. 1990. "An Orientation toward Help-Seeking for Emotional Problems." *Social Science and Medicine* 31: 989-995.

Finney, J. W., R. E. Mitchell, R. C. Cronkite et al. 1984. "Methodological Issues in Estimating Main and Interactive Effects: Examples from Coping Social Support and Stress Field." *Journal of Health and Social Behavior* 25: 85-98.

Fisher, G., P. Benson, and R. Tessler. 1990. "Family Response to Mental Illness: Developments Since Deinstitutionalization." Pp. 203-236 in *Research in Community Mental Health,* vol. 6, edited by J. Greenley. Greenwich, CT: JAI Press.

Fortney, J., K. Rost, and M. Zhang. 1998. "A Joint Choice Model of the Decision to Seek Depression Treatment and Choice of Provider Sector." *Medical Care* 36: 307-320.

Fosu, G. B. 1995. "Women's Orientation Toward Help-Seeking for Mental Disorders." *Social Science and Medicine* 40: 1029-1040.

Fox, J. 1984. "Sex, Marital Status, and Age as Social Selection Factors in Recent Psychiatric Treatment." *Journal of Health and Social Behavior* 25: 394-405.

Gallo, J. J., S. Marino, D. E. Ford et al. 1995. "Filters on the Pathway to Mental Health Care, II. Sociodemographic Factors." *Psychological Medicine* 25: 1149-1160.

Glass, T. A., C. F. Mendes de Leon, T. E. Seeman et al. 1997. "Beyond Single Indicators of Social Networks: A Lisrel Analysis of Social Ties among the Elderly." *Social Science and Medicine* 44: 1503-1517.

Goffman, E. 1961. *Asylums.* New York: Anchor.

Goldberg, D. 1972. *The Detection of Psychiatric Illness by Questionnaire.* London: Oxford University Press.

Golding, J., and K. B. Wells. 1990. "Social Support and Use of Mental Health Services by Mexican Americans and Non-Hispanic Whites." *Basic and Applied Social Psychology* 11: 443-458.

Gore, S. L. 1978. "The Effect of Social Support in Moderating the Health Consequences of Unemployment." *Journal of Health and Social Behavior* 19: 157-165.

Gove, W., and J. F. Tudor. 1973. "Adult Sex Roles and Mental Illness." *American Journal of Sociology* 98: 812-835.

Greenley, J. R., and D. Mechanic. 1976. "Social Selection in Help Seeking for Psychological Problems." *Journal of Health and Social Behavior* 17: 249-262.

Greenley, J. R., D. Mechanic, and P. D. Cleary. 1987. "Seeking Help for Psychologic Problems: A Replication and Extension." *Medical Care* 25: 1113-1128.

Henderson, S., P. Duncan-Jones, D. G. Byrne et al. 1980. "Measuring Social Relationships: The Interview Schedule for Social Interaction." *Psychological Medicine* 10: 723-734.

Horgan, C. 1986. "Specialty and General Ambulatory Mental Health Services: Comparison of Utilization and Expenditures." *Archives of General Psychiatry* 42: 565-572.

Horwitz, A. 1977. "Social Networks and Pathways to Psychiatric Treatment." *Social Forces* 56: 86-105.

Horwitz, A. V., and S. C. Reinhard. 1995. "Family and Social Network Supports for the Seriously Mentally Ill: Patient Perceptions." *Research in Community and Mental Health* 8: 205-232.

Horwitz, A. V. 1992. "Adult Siblings as Sources of Social Support for the Seriously Mentally Ill: A Test of the Serial Model." *Journal of Marriage and the Family* 55: 623-632.

Horwitz, A. V., and S. C. Reinhard. 1995. "Family and Social Network Supports for the Seriously Mentally Ill: Patient Perceptions." *Research in Community and Mental Health* 8: 205-232.

Horwitz, S. M., H. Morgenstern, and L. F. Berkman. 1985. "The Impact of Social Stressors and Social Networks on Pediatric Medical Care Use." *Medical Care* 23: 946-959.

Hough, R. L., J. A. Landsverk, M. Karno et al. 1978. "Utilization of Health and Mental Health Services by Los Angeles Mexican Americans and Non-Hispanic Whites." *Archives of General Psychiatry* 44: 702-709.

House, J. S. 1981. *Work Stress and Social Support.* Reading, MA: Addison-Wesley.

House, J. S., and R. L. Kahn. 1985. "Measures and Concepts of Social Support." Pp. 83-108 in *Social Support and Health,* edited by S. Cohen. New York: Academic Press.

House, J. S., C. A. Robbins, and H. Metzner. 1982. "The Association of Social Relationships and Activities with Mortality: Prospective Evidence from the Tecumseh Community Health Study." *American Journal of Epidemiology* 116: 123-140.

House, J. S., C. Umberson, and K. R. Landis. 1988. "Structures and Processes of Social Support." *Annual Review of Sociology* 14: 293-318.

Howard, K. I., T. A. Cornille, J. S. Lyons et al. 1996. "Patterns of Mental Health Service Utilization." *Archives of General Psychiatry* 53: 696-703.

Kadushin, C. 1969.*Why People Go to Psychiatrists.* New York: Atherton.

Kaplan, B. H., J. C. Cassell, and S. Gore. 1977. "Social Support and Health." *MedicalCare* 15: 47-58.

Kasl, S., S. Gore, and S. Cobb. 1975. "The Experience of Losing a Job: Reported Changes in Health Symptoms and Illness Behaviors." *Psychological Medicine* 37: 106-121.

Katz, S. J., R. C. Kessler, R. G. Frank et al. 1997. "Mental Health Care Use, Morbidity, and Socioeconomic Status in the United States and Ontario. *Inquiry* 34: 38-49.

Keitner, G. I., C. E. Ryan, I. W. Miller et al. 1995. "Role of the Family in Recovery and Major Depression." *American Journal of Psychiatry* 152(7): 1002-1008.

Kessler, L. G., R. Folsom, R. Royall et al. 1985. "Parameter and Variance Estimation." Pp. 327-350 in *Epidemiologic Field Methods in Psychiatry: The NIMH Epidemiologic Catchment Area Program*, edited by W. W. Eaton and L.G. Kessler. Orlando, FL: Academic Press.

Kessler, R. C., K. A. McGonage, S. Zhao et al. 1994. "Lifetime and 12-month Prevalence of DSM-III-R Psychiatric Disorders in the United States: Results from the National Comorbidity Survey." *Archives of General Psychiatry* 51: 8-19.

Kessler, R. C., and J. D. McLeod. 1985. "Social Support and Mental Health in Community Samples." Pp. 19-49 in *Social Support and Health*, edited by S. Cohen. New York: Academic Press.

Kessler, R. C., S. Zhao, S. Katz, A. Kouzis et al. 1999. "Past Year Use of Outpatient Services for Psychiatric Problems in the National Comorbidity Survey." *American Journal of Psychiatry* 156: 115-123.

Kiesler, C. A. 1985. "Policy Implications of Research on Social Support and Health." Pp. 347-164 in *Social Support and Health*, edited by S. Cohen and S. L. Syme. Orlando, FL: Academic Press.

Kouzis, A. C., and W. W. Eaton. 1998. "Absence of Social Support, Social Networks and Health Services Utilization." *Psychological Medicine* 28: 1301-1310.

Kramer, M. 1983. "The Continuing Challenge: The Rising Prevalence of Mental Disorders, Associated Chronic Diseases and Disabling Conditions." *The American Journal of Social Psychiatry* 3: 13-24.

Krause, N. 1988. "Stressful Life Events and Physician Utilization." *Journal of Gerontology* 43: S53-61.

Landerman, L.R., B. J. Burns, M. S. Swartz, H. R. Wagner, and L. K. George. 1994. "The Relationship Between Insurance Coverage and Psychiatric Disorder in Predicting Use of Mental Health Services." *American Journal of Psychiatry* 151: 1785-1790.

Laireiter, A., and U. Baumann. 1992. "Network Structures and Support Functions: Theoretical and Empirical Analyses." Pp. 57-62 in *The Meaning and Measurement of Social Support*, edited by H. O. F. Veiel and U. Baumann. New York: Hemisphere.

Langner, T., and S. Michael. 1963.*Life Stress and Mental Health.* New York: Free Press.

Leaf, P. J., M. M. Livingston, G. L. Tischler et al. 1985."Contact with Health Professionals for the Treatment of Psychiatric and Emotional Problems." *Medical Care* 23: 1322-1337.

Leaf, P. J., and M. L. Bruce. 1987."Gender Differences in the Use of Mental Health-related Services: A Re-examination." *Journal of Health and Social Behavior* 23: 171-183.

Leaf, P. J., M. L. Bruce, G. L. Tischler et al. 1988. "Factors Affecting the Utilization of Specialty and General Medical Health Services." *Medical Care* 26: 9-26.

Lefley, H. P. 1996. *Family Care Giving in Mental Illness.* Newbury Park, CA: Russell Sage.

Lin, N., and K. Peek. 1999. "Social Networks and Mental Health." Pp. 241-258 in *A Handbook for the Study of Mental Health*, edited by A. Horwitz and T. L. Scheid. Cambridge: Cambridge University Press.

Lin N., R. Simeone, W. Ensel, and W. Kuo. 1979. "Social Support, Stressful Life Events, and Illness: A Model and an Empirical Test." *Journal of Health and Social Behavior* 20: 108-119.

Lin, N., M. W. Woelfel, and S. C. Light. 1985. "The Buffering Effect of Social Support Subsequent to an Important Life Event." *Journal of Health and Social Behavior* 26: 247-263.

Link, B. 1987. "Understanding Labeling Effects in the Area of Mental Illness: An Assessment of the Effects of Expectations of Rejection." *American Journal of Sociology* 52: 96-112.

Link, B. G., and B. P. Dohrenwend. 1980. "Formulation of Hypotheses about the Ratio of Untreated and Treated Cases in the True Prevalence Studies of Functional Disorders in Adults in the United States." Pp. 133-149 in *Mental Illness in the United States: Epidemiological Estimates*, edited by B. P. Dohrenwend, B. S. Dohrenwend, M. S. Gould et al. New York: Praegar.

Mays, V. M., C. H. Caldwell, and J. S. Jackson. 1996. "Mental Health Symptoms and Service Utilization Patterns of Help-seeking among African-American Women." Pp. 161-176 in *Mental Health in Black America*, edited by H. W. Neighbors and J. S. Jackson. Thousand Oaks, CA: Sage Publications.

Mechanic, D. 1979. "Correlates of Physician Utilization: Why Do Major Multivariate Studies of Physician Utilization Find Trivial Psychosocial and Organizational Effects?" *Journal of Health and Social Behavior* 20: 387-396.

Mechanic, D., R. J. Angel, and L. Davies. 1992. "Correlates of Using Mental Health Services: Implications of Using Alternative Definitions." *American Journal of Public Health* 82: 74-78.

Mirowsky, J., and C. E. Ross. 1989. *Social Causes of Psychological Distress*. New York: Aldine de Gruyter.

Morrissey, J., and H. Goldman. 1986. "Care and Treatment of the Mentally Ill in the United States: Historical Developments and Reforms." *Annals of the American Academy of Political and Social Science* 484: 12-27.

Narrow, W., D. A. Regier, D. S. Rae et al. 1993. "Use of Services by Persons with Mental Addictive Disorders: Findings from the NIMH ECA Program." *Archives of General Psychiatry* 50: 95-107.

Neighbors, H. W., and J. S. Jackson. 1996. "Mental Health in Black America: Psychosocial Problems and Help-seeking Behavior." Pp. 1-13 in *Mental Health in Black America*, edited by H. W. Neighbors and J. S. Jackson. Thousand Oaks, CA: Sage Publishers.

Nuckolls, K. B., J. C. Cassell, and B. H. Kaplan. 1972. "Psycho-social Assets, Life Crises, and the Prognosis of Pregnancy." *American Journal of Epidemiology* 95: 431-441.

Olfson, M., and G. L. Klerman. 1992. "Depressive Symptoms and Mental Health Service Utilization in a Community Sample." *Social Psychiatry and Psychiatric Epidemiology* 27: 161-167.

Olfson, M., and H. A. Pincus. 1994. "Outpatient Psychotherapy in the United States, I: Volume, Costs, and User Characteristics." *American Journal of Psychiatry* 151: 1281-1288.

Padgett, D. K., C. Patrick, B. J. Burns et al. 1994. "Ethnicity and the Use of Outpatient Mental Health Services in a National Insured Population." *American Journal of Public Health* 84: 222-226.

Pearlin, L. I. 1999. "Stress and Mental Health: A Conceptual Overview." Pp. 161-175 in *A Handbook for the Study of Mental Health: Social Contexts, Theories, and Systems*, edited by A. V. Horwitz and T. L. Scheid. Cambridge: Cambridge University Press.

Pearlin, L. I., and C. Schooler. 1978. "The Structure of Coping." *Journal of Health and Social Behavior* 19: 2-21.

Pescosolido, B. A. 1992. "Beyond Rational Choice: The Social Dynamics of How People Seek Help." *American Journal of Sociology* 97: 1096-1138.

Pescosolido, B. A. 1996."Bringing the 'Community' into Utilization Models: How Social Networks Link Individuals to Changing Systems of Care." *Research in the Sociology of Health Care* 13A: 171-197.

Pescosolido, B. A., and J. J. Kronenfeld. 1995. "Health, Illness and Healing in an Uncertain Era: Challenges from and for Medical Sociology." *Journal of Health and Social Behavior* 36 (extra issue): 5-33.

Pescosolido, B. A., E. R. Wright, M. Alegria et al. 1998. "Social Networks and Patterns of Use among the Poor with Mental Health Problems in Puerto Rico." *Medical Care* 36: 1057-1072.

Pescosolido, B. A., and C. A. Boyer. 1999. "How Do People Come to Use Mental Health Services? Current Knowledge and Changing Perspectives." Pp. 392-411 in *A Handbook for the Study of Mental Health*, edited by A. V. Horwitz and T. L. Scheid. Cambridge: Cambridge University Press.

Pilisuk, M., R. Boyland, and C. Acredolo. 1987. "Social Support, Life Stress, and Subsequent Medical Care Utilization." *Health Psychology* 6: 273-288.

Pincus, H. A., D. A. Zarin, T. L. Tanielian et al. 1999. "Psychiatric Patients and Treatments in 1997: Findings from the American Psychiatric Practice Research Network." *Archives of General Psychiatry* 56: 441-449.

Portes, A., Keyle D., and W. W. Eaton. 1992. "Mental Illness and Help-seeking Behavior among Mariel Cuban and Haitian Refugees in South Florida." *Journal of Health and Social Behavior* 33: 283-298.

Regier, D. A. 1982. "Specialist/Generalist Division of Responsibility for Patients with Mental Disorders." *Archives of General Psychiatry* 39: 219-224.

Regier, D. A., I. D. Goldberg, and C. A. Taube. 1978."The De Facto U.S. Mental Health Services Systems: A Public Health Perspective." *Archives of General Psychiatry* 35: 695-698.

Regier, D. A., C. T. Kaelber, D. S. Rae et al. 1998. "Limitations of Diagnostic Criteria and Assessment Instruments for Mental Disorders: Implications for Research and Policy." *Archives of General Psychiatry* 55:109-115.

Regier, D. A., W. E. Narrow, D. S. Rae et al. 1993. "The De Facto U.S. Mental Health and Addictive Disorders Service System: Epidemiologic Catchment Area Prospective 1-Year Prevalence Rates of Disorders and Services." *Archives of General Psychiatry* 50: 85-94.

Roberts, B. L., R. Dunkle, and M. Haug. 1994. "Physical, Psychological, and Social Resources as Moderators of the Relationship of Stress to Mental Health of the Very Old." *Journal of Gerontology* 49: 35-43.

Robins, L. N., and D. A. Regier (Eds.). 1991. *Psychiatric Disorders in America: The Epidemiologic Catchment Area Study*. New York: Free Press.

Rook, K. S. 1984. "Promoting Social Bonding: Strategies for Helping the Lonely and Socially Isolated." *American Psychologist* 39: 1389-1407.

Rosenfield, S. 1997. "Labeling Mental Illness: The Effects of Received Services and Perceived Stigma on Life Satisfaction." *American Sociological Review* 62: 660-672.

Ross, C. E. 1995. "Reconceptualizing Marital Status as a Continuum of Social Attachment." *Journal of Marriage and the Family* 57: 129-140.

Rossi, A., and P. Rossi. 1990. *Of Human Bonding: Parent-Child Relations across the Life Course*. New York: Aldine de Gruyter.

Seeman, T., and L. F. Berkman. 1988. "Structural Characteristics of Social Networks and Their Relationship with Social Support in the Elderly: Who Promotes Support?" *Social Science and Medicine* 26: 737-749.

Shapiro, S., E. A. Skinner, P. S. German et al. 1986. "Need and Demand for Mental Health Services in an Urban Community: An Exploration Based on Household Interviews." Pp. 307-320 in *Mental Disorders in the Community: Progress and Challenge*, edited by J. Barrett. New York: Guilford Press.

Shapiro, S., E. A. Skinner, L. G. Kessler et al. 1984. "Utilization of Health and Mental Health Services: Three Epidemiologic Catchment Area Sites." *Archives of General Psychiatry* 41: 971-978.

Shapiro, S., G. L. Tischler, and L. Cottler et al. 1985. "Health Services Research Questions." Pp. 191-208 in *Epidemiologic Field Methods in Psychiatry: The NIMH Epidemiologic Catchment Area Program*, edited by W. W. Eaton and L.G. Kessler. Orlando, FL: Academic.

Sherbourne, C. D. 1988. "The Role of Social Support and Life Stress Events in Use of Mental Health Services." *Social Science Medicine* 27: 1393-1400.

Stronks, K., H. van de Mheen, C. W. N. Looman et al. 1998. "The Importance of Psychosocial Stressors for Socio-economic Inequalities in Perceived Health." *Social Science and Medicine* 46: 611-623.

Sturm, R., C. A. Jackson, L. S. Meredith et al. 1995. "Mental Health Utilization in Prepaid and Fee-for-Service Plans among Depressed Patients in the Medical Outcomes Study." *Health Services Research* 30: 319-340.

Sussman, L. K., L. N. Robins, and F. Earls. 1987. "Treatment-seeking for Depression by Black and White Americans." *Social Science and Medicine* 3: 187-196.

Sutton, J. R. 1991. "The Political Economy of Madness: The Expansion of the Asylum in Progressive America." *American Sociological Review* 56: 665-678.

Thoits, P. A. 1982. "Conceptual, Methodological, and Theoretical Problems in Studying Social Support as a Buffer Against Life Stress." *Journal of Health and Social Behavior* 23: 145-159.

Thoits, P. A. 1992. "Social Support Functions and Network Structures: A Supplemental View." Pp. 57-62 in *The Meaning and Measurement of Social Support*, edited by H. O. F. Veiel and U. Baumann. New York: Hemisphere.

Turner, R. J. 1981. "Social Support as a Contingency in Psychological Well-being." *Journal of Health and Social Behavior* 22: 357-367.

Turner, R. J., and F. Marino. 1994. "Social Structure and Social Support: A Descriptive Epidemiology." *Journal of Health and Social Behavior* 35: 193-212.

Tyrer, P., S. Merson, S. Onyett et al. 1994. "The Effect of Personality Disorder on Clinical Outcome, Social Networks and Adjustment: A Controlled Clinical Trial of Psychiatric Emergencies." *Psychological Medicine* 24: 731-740.

Vaux, A. 1988. *Social Support: Theory, Research, and Intervention.* New York: Praeger.

Verhaak, P. F. M. 1995. "Determinants of the Help-seeking Process: Goldberg and Huxley's First Level and First Filter." *Psychological Medicine* 25: 95-104.

Von Korff, M., L. Cottler, L. K. George et al. 1985. "Nonresponse and Nonresponse Bias in the ECA Surveys." Pp. 85-98 in *Epidemiologic Field Methods in Psychiatry: The NIMH Epidemiologic Catchment Area Program*, edited by W. W. Eaton and L. G. Kessler. Orlando, FL: Academic Press.

Waite, L. J. 1995. "Does Marriage Matter?" *Demography* 32: 483-501.

Wells, K. B., A. Burnam, and P. Camp. 1995. "Severity of Depression in Prepaid and Fee-for-Service General Medical and Mental Health Specialty Practices." *Medical Care* 33: 350-364.

Wells, K. B., J. M. Golding, R. L. Hough et al. 1988. "Factors Affecting the Probability of Use of General and Medical Health and Social/Community Services for Mexican Americans and Non-Hispanic Whites." *Medical Care* 26: 441-452.

Wells, K. B., W. Manning, and B. Benjamin. 1986. "A Comparison of the Effects of Sociodemographic Factors and Health Status on Use of Outpatient Mental Health Services in HMO and Fee-for-Service Plans." *Medical Care* 24: 949-960.

Wells, K. B., W. G. Manning, Duan N. et al. 1982. *Cost Sharing and Demand for Ambulatory Mental Health Services.* Santa Monica, CA: Rand, DHHS Publication No. 2960-HHS.

Wells, K. B., W. G. Manning, N. Duan et al. 1986. "Sociodemographic Factors and the Use of Outpatient Mental Health Services." *Medical Care* 24: 75-85.

Williams, A. W., J. E. Ware, Jr., and C. A. Donald. 1981. "A Model of Mental Health, Life Events, and Social Supports Applicable to General Populations." *Journal of Health and Social Behavior* 22: 324-336.

Wolinsky, F. D. 1976. "Health Services Utilization and Attitudes toward Health Maintenance Organizations: A Theoretical and Methodological Discussion." *Journal of Health and Social Behavior* 17: 221-236.

Wolinsky, F., and R. J. Johnson. 1991. "The Use of Health Services by Older Adults." *Journal of Gerontology* 46: S345-357.

Wu, L. T., A. C. Kouzis, and P. J. Leaf. 1999. "Influence of Comorbid Alcohol and Psychiatric Disorders on Utilization of Mental Health Services in the National Comorbidity Survey." *American Journal of Psychiatry* 156: 1230-1236.

PART II

TREATMENT AND RECOVERY PROCESSES

A REVIEW OF EMPIRICAL STUDIES OF INTERVENTIONS FOR FAMILIES OF PERSONS WITH MENTAL ILLNESS

David E. Biegel, Elizabeth A. R. Robinson, and
Marilyn J. Kennedy

ABSTRACT

This paper examines the effects on families of a broad range of intervention programs designed to assist family caregivers of persons with mental illness. In so doing, the paper critically discusses the intervention designs utilized, similarities and differences in intervention modalities, the characteristics of study subjects, and the effects of these intervention programs on a variety of caregiver outcomes. The paper addresses the degree to which particular interventions are more effective than others, and assesses the strengths, weaknesses, and limitations of these intervention studies. Implications for future practice, policy, and research are presented.

Research in Community and Mental Health, Volume 11, pages 87-130.
Copyright © 2000 by JAI Press Inc.
All rights of reproduction in any form reserved.
ISBN: 0-7623-0671-8

INTRODUCTION

Chronic illness, including mental illness, has significant effects on patients, families, and the larger society. Providing care to a chronically ill person affects family caregivers, those family members who provide the most support and assistance to their ill family member. Across chronic illnesses, many family caregivers report experiencing moderate to high levels of burden and some caregivers experience moderate to high levels of depression as well (Biegel, Sales, and Schulz 1991; Bruhn 1977).

Studies of families with a member who has mental illness indicate that family burdens of caregiving are persistent and pervasive. These family caregivers suffer from a number of significant stresses, experience moderately high levels of burden, and often receive inadequate assistance from mental health professionals (Biegel, Song, and Milligan 1995; Davis, Dinitz, and Pasaminick 1974; Grad and Sainsbury 1963; Hatfield 1978; Holden and Lewine 1982; Kreisman and Joy 1974; Lefley 1996).

As with other chronic illnesses, a number of different types of interventions have been developed to address the needs of families with a member who has mental illness. Some of the earliest interventions, for example psychoeducational interventions, were primarily focused on patient outcomes, that is reducing rates of patient recidivism, rather than on caregiver needs. These family interventions attempted to change caregiver behaviors (i.e., high expressed emotion) associated with high patient relapse rates (e.g., Tarrier et al. 1988). Other early interventions, such as self-help groups for families of the mentally ill, were developed by families independent of mental health professionals and mental health agencies who were felt to be unsupportive of family needs (Hatfield 1986). Over time, all these interventions have matured and changed such that psychoeducational interventions now address family burden as well as patient recidivism, and support groups are often sponsored by mental health agencies with mental health professionals acting as consultants (Biegel, Sales, and Schulz 1991). In addition, a range of other types of interventions, such as educational programs, counseling, respite care, and individual family consultation have been developed to address the needs of families of persons with mental illness.

There have been a number of literature reviews which have examined the effects of these programs. However, these reviews have either focused primarily on patient rather than family outcomes (e.g., Barrowclough and Tarrier 1984; Lam 1991; Mari and Streiner 1994; Schooler, Kith, Severe, and Matthews 1995) or have examined only particular types of interventions (Dixon and Lehman 1995; Penn and Mueser 1996). One broader review of interventions for family caregivers of five different chronic illnesses has been published (Biegel, Sales, and Schulz 1991). However, the material in this review is somewhat dated, and due to its cross-illness scope, there is insufficient depth in the section on mental illness. As of yet, there has been no up to date, comprehensive, in-depth literature review which has examined the full range of intervention modalities for family caregivers of persons with mental illness.

The purpose of this paper is to examine the effects on families of a broad range of intervention programs designed to assist family caregivers of persons with mental illnesses. In doing so, we will examine and critically discuss the intervention designs utilized, similarities and differences in intervention modalities, the characteristics of study subjects, and the effects of interventions on a variety of caregiver outcomes. The paper will address the degree to which particular interventions are more effective than others, examine which types of outcomes are more likely to occur than others, and assess the strengths, weaknesses, and limitations of these intervention studies. Implications for future practice, policy, and research will be presented.

METHODOLOGY

Studies for this review were selected on the basis of two general criteria: each study must contain (1) an intervention specifically for families with an adult mentally ill member(s) and (2) at least one family outcome measure. Intervention studies which used solely client outcomes were omitted from the review; however, those which contained both family and client outcomes are included here, with principal emphasis given to the former findings. In addition, studies that included only qualitative findings or only measures of satisfaction with the intervention were excluded from this review. Further, the determination was made to consider as family measures only those variables which had as their primary target, family rather than client functioning. Thus, studies that limited themselves to such outcome measures as reduction in expressed emotion (EE) were judged to have the client's functioning as their primary objective and were, therefore, excluded. In contrast, variables directly related to functioning of the family unit, such as reduction in overall family conflict, overall changes in affective climate or family attitude, increased problem-solving and behavior management skills, acquisition of knowledge-of-illness by family members, and family burden are included in this analysis. There was, at times, a fine line between inclusion and exclusion of some variables, and ultimately of some studies. In those instances, the perceived intent or thrust of the study became the determining factor.

Sixty-four articles representing 43 research projects, published during a 17-year time span, 1980 to 1997, met the foregoing criteria and are included in this review. They were located by systematic online searches of eight data bases[1] from 1980 through 1997, careful examination of references and bibliographies contained in recent books and articles, and inspection of recent review articles in related areas, many of which had a more limited focus. Further, in an attempt to include current projects which might not yet be electronically catalogued, visual scanning of recent journal issues was conducted. In a few instances personal communication with principal investigators of projects known to be in process was initiated, and some additional material as well as cross-references to other projects were obtained from these contacts.

The intervention studies reported here are organized into four broad categories as described below:

(1) *Support Groups.* A group-based intervention intended to provide families and family caregivers with social and psychological support, and in most instances with varying degrees of information about mental illness, as well as resources for coping with that illness. Support groups may be professionally led (usually time-limited) or peer led (usually open-ended and ongoing). The former are often conducted in facilities such as hospitals or clinics, whereas the latter are typically held in community meeting places such as churches, libraries, or social centers.

(2) *Education.* A group-based intervention for families of persons with mental illness, usually of short duration (less than six months) focusing on provision of cognitive information on mental illness and its management, including advice for family members. This intervention may either be led by mental health professionals or by trained, lay group leaders.

(3) *Psychoeducation.* An intervention for families and family members with mental illness, led by a mental health professional or a team of mental health professionals. Psychoeducational interventions have, as a principal component, a focus on changing family coping behaviors and attitudes. In addition to the presentation of cognitive information, they typically include training families in communication and problem-solving. Although the designated target of change is the family, the ultimate impact of psychoeducation is on the patient's symptoms and functioning. Psychoeducational interventions are provided in both multi-family and single family formats.

(4) *Individual Family Support and Multi-Component Interventions.* Individual family support is an individual family intervention in which counseling, therapy, and/or other family support services are provided directly to individual family units or their members by mental health professionals. This category consists of a wide variety of services, the most prominent of which are family therapy, individual supportive therapy, family consultation, and case management. The majority of these services are provided on an open-ended, as-needed basis, with the option of becoming long term. Multi-component interventions are state-wide family support and education programs in two states (Massachusetts and New Jersey). Services offered by the programs in one or both of these two states are very broad and include individual family support as well as support groups, education, and psychoeducation. In addition, they offer respite care, advocacy, patient companion program, and technical assistance to family support groups (Benson et al. 1996; Riesser, Minsky, and Schorske 1991).

Table 1. Support Group Interventions for Families Interventions, Research Designs, and Subjects

Study	Interventions	Research Design[*]	Subjects (N = *Individuals, Not Families*)
Anderson et al. (1986)	Three 4-hr meetings of a) professionally led support group or b) psychoeducational group (see Table 3)	Experimental: (2 conditions with pre & post data); plus a post-hoc no-treatment comparison group	N = 40 patients (20 per condition), all with affective disorders; primarily female, mean age = 50. Families mostly spouses, some parents.
Biegel & Yamatani (1986;1987)	Ongoing member-led support groups	Quasi-experimental: (one condition with pre & post data at 6 mo)	N = 45 family caregivers. Predominately female, Caucasian, parents
Heller et. al. (1997)	Ongoing support groups; over 2/3 led by group members	Non-experimental: (survey of support group members)	N = 131 family caregivers. Predominately female, Caucasian, some college, average age = 57. Consumers predominately male, with schizophrenia, average age = 37.
Kane et al. (1990)	Four week (2 hrs/wk) of professionally led: a) education group or b) support group	Quasi-experimental: (non-random assignment to 2 conditions with pre & post data)	N = 49 (25 in support group and 24 in education group) family caregivers. Primarily Caucasian, parents, majority higher social class.
Medvene & Krauss (1989)	Ongoing member-led support group	Non-experimental: (survey of support group members)	N = 57 parents, including 36 couples. Predominately female, Caucasian, college-educated and middle-aged. Consumers predominately male, young adults, high school educated.hospitalized at least twice.

(continued)

Table 1 (Continued)

Study	Interventions	Research Design[*]	Subjects (N = individuals, not families)
Medvene et al. (1995)	Ongoing support groups led by bilingual Hispanic mental health professionals.	Quasi-experimental: (one condition with pre & post data 12 mo.)	N = 32 Mexican-American parents. Predominately female, over 60 yrs. of age, elementary school education, low family income. Consumers predominately male, 10th grade education, mean age = 33.
Mills & Hansen (1991)	5-week groups, 3 hrs/wk, in residential facility: a) psychoeducational groups or b) support groups, both professionally led.	Experimental: (2 intervention conditions with pre, post & 3-month follow-up data; Note: Ss/conditions not equivalent at baseline)	N = 12 patients with schizophrenia in residential facility. Predominantly male, all white, 19-40 yrs old. Families all parents, 40-70 yrs old.
Monking (1994)	On-going family support groups for participants in a one-year therapeutic intervention	Quasi-experimental: (intervention & matched comparison conditions with pre & post data)	N = 19 caregivers with one year of support group participation. German study.
Norton et al. (1993)	On-going support groups affiliated with the National Alliance of the Mentally Ill	Non-experimental: (survey of support group members and non-support group family members)	N = 99 (40 NAMI group and 59 non-NAMI members). NAMI members predominately Caucasian, highly educated. higher social class; non-members predominantly African-American, less well educated, lower social class.
Potasznik & Nelson (1984)	Three ongoing support groups	Quasi-experimental: (one condition with post data only)	N = 56 parents; predominately female. Patients predominately male. Canadian study.

Note: * All experimental designs used random assigment to conditions.

It should be noted that there is inevitably some overlap among categories, particularly between education and support groups. As noted earlier, most support groups do contain an element of education, although usually less formal in nature, and many education groups attempt to provide social support for their participants. The deciding factor lies in the principal focus of the intervention.

INTERVENTION DESCRIPTIONS, RESEARCH DESIGNS, AND SUBJECTS

Our review of these 43 intervention studies begins with a description of their methodology. After a brief description of the interventions in each category, we review the research designs, samples, and the outcomes measured. Following this, the paper describes the findings of the studies, again presented by category of family intervention.

It should be noted that studies often provided very little information about the independent variable, the intervention(s) tested. Thus, while we were able to classify intervention studies into the four broad categories analyzed in this paper, for many studies the theoretical underpinning of the intervention as well as the specific intervention components that were delivered remain somewhat unclear.

Intervention Descriptions

Tables 1 through 4 provide details on the methodology of the studies in the four intervention categories, including some details on the interventions themselves. Table 1 presents this information for the ten studies of support group interventions. Six of these studies investigated ongoing support groups, whereas the other four studies were of time-limited groups, specifically developed for the research study. Only one ongoing support group was professionally led; the rest were all member led. All the time-limited support groups were professionally led. The duration of support groups ranged from four hours on one day to 15 hours over a five-week period.

Table 2 presents information on the methodology of the seventeen educational interventions. The format of these interventions was generally didactic and structured around the presentation of content; this is particularly true of the shorter interventions. About half of the programs provided opportunities for informal group discussion and in several the provision of social support was integral to the presentation of content. Most of these interventions were presented in small group sessions in which social support between members would have probably occurred. Though the length of these interventions varied greatly, most typical are weekly 1.5-hour sessions over two to 16 weeks.

Table 2. Educational Interventions for Families Interventions, Research Designs, and Subjects

Study	Interventions	Research Design[*]	Subjects (N = Individuals, not Families)
Abramowitz & Coursey (1989)	6 week education & support group	Experimental: (intervention & control conditions with pre & post data)	N = 48 (24 per condition). Predominantly mothers, mean age = 51, majority white. Majority of patients male; ages 25-35, all with schizophrenia.
Barrowclough et al. (1987)	2 individual family sessions; used booklet & talk by professional; patient included in 2nd session.	Quasi-experimental: (one condition with pre & post data)	N = 24. British study. Majority were mothers with some fathers & spouses; predominantly high EE. Majority of patients female, mean age = 37, 6.6 yrs since onset, 3.2 prior hospitalizations, all with schizophrenia.
Berkowitz et al. (1984)	2-4 talks in family's home	Experimental: (intervention & control conditions with pre, post & 9-month follow-up data)	N = 36 (intervention condition = 21; control = 15). British study. Equal numbers of parents & spouses; majority high EE & male. Majority of patients male, mean age = 38, age of onset = 30, 2.4 prior hospitalizations, all with schizophrenia.
Birchwood, Smith & Cochrane (1992)	Similar content in: 4 small group sessions, by mail, or in video tapes; *and* x) with or y) without homework	Quasi-experimental: (non-random assign-ment to 6 inter-vention conditions with pre, post & 6-month follow-up data)	N = 94 (group intervention = 47; mail = 30; video = 17; about half received homework assignments & half did not). British study. Family living with or in close contact with patient; excluded those scoring high on knowledge; patients predominantly male, mean age = 30, 8 yrs since onset, 2.7 prior hospitalizations, all with schizophrenia.
Canive et al. (1996)	6 weekly classes, called "psychoeducation," primarily information.	Quasi-experimental: (one condition with pre, post & 9-month follow-up data)	N = 68 (41 mothers, 27 fathers). Spanish study.
Cazullo et al. (1989)	16-week didactic & discussion groups	Quasi-experimental (one condition with pre, post & 9-month follow-up data)	N = 14. Italian study. Pedominantly mothers from two-parent families. All patients diagnosed with schizophrenia.
Ehlert (1988)	a) 10-session education & problem solving groups, b) 5-session education only groups, or c) 5-session problem-solving only group.	Experimental: (3 conditions with pre, post & 9-month follow-up data)	N = 49 (N in each condition unclear). German study. Predominantly mothers, mean age = 52, predominantly married. Majority of patients male, 7 yrs since onset, mean age = 30, all with schizophrenia.

Citation	Intervention	Design	Sample
Glynn, Pugh & Rose (1993); Glynn, Pugh & Rose (1990)	Day-long didactic group workshop	Quasi-experimental: (one condition with pre, post & 3-month follow-up data)	N = 33. Majority mothers, mean age = 52, majority white, some Hispanic. Majority of patients male, mean age = 27, mean age of onset = 18, all with schizophrenia or schizo-affective disorder and currently hospitalized.
Hugen (1993)	Day-long didactic group workshop	Quasi-experimental: (one condition with pre, post & 3-month follow-up data)	N = 22. Family members' mean age = 52, majority parents, all but one white, predominantly female. Patient mean age = 48, mean age of onset = 18, almost equal males & females, all with schizophrenia.
Kane, DiMartino & Jimenez (1990)	Four week (2 hrs/ wk) of professionally led: a) education group or b) support group	Quasi-experimental: (non-random assign-ment to 2 conditions with pre & post data)	N = 49 (education group = 18; support group = 31). Predominantly lower social class, white, & mothers with some fathers, mean age = 49. Mean patient age = 26, 6 yrs since onset, 4 prior hospitalization; all with schizophrenia or schizoaffective disorder.
Kuipers et al. (1989); MacCarthy et al. (1989)	Year-long monthly education groups with counseling & support.	Quasi-experimental: (non-random assign-ment to intervention & control conditions with pre and 9 or 12 month follow-up)	N = 30 (education = 13; control = 17). British study. Half of family members were mothers. Mean patient age = 36; majority men; 15-20 yrs since onset; half with schizophrenia, some bipolar disorders; all attended a hospital day program.
Mannion, Mueser & Solomon (1994)	12-week educational groups for spouses	Quasi-experimental: (one condition with pre, post & 1-year follow-up data)	N =19 spouses. Almost all white, 74% female, mean length of marriage = 19 yrs. Majority of patients bipolar disorder, 36% had schizophrenia or schizoaffective disorder; mean of 13 yrs since onset.
Orhagen & d'Elia (1992 a & b)	Six-session education & discussion groups	Quasi-experimental: (one condition with pre & post data)	N = 47. Swedish study. Predominantly parents (mothers & fathers). Patients predominantly male, mean age = 35, 12 yrs since onset, 9 prior hospitalizations, all with schizophrenia.

(continued)

Table 2 (Continued)

Study	Interventions	Research Design[*]	Subjects (N = Individuals, not Families)
Posner et al. (1992)	Eight-week education & support groups	Experimental: (intervention & control conditions with pre, post & 6-month follow-up data)	N = 55 family members (N per condition un-clear). Half were mothers with some fathers, spouses & siblings; majority lived with patient. Patients predominantly men, mean age = 29, 4.4 prior hospitalizations; all with schizophrenia.
Sidley, Smith & Howells (1991)	Two 90-min. group education sessions; content varied by condition.	Quasi-experimental: (non-random assignment to 2 conditions with pre & post data)	N = 18 (N in each condition unclear). British study. Predominantly mothers, with some fathers & spouses. Patients predominantly male, mean age = 34, 12 yrs since onset, 6 prior hospitalizations, all with schizophrenia.
Smith & Birchwood (1987)	a) 4-week education groups or b) mailing booklets to families	Experimental: (2 intervention conditions with pre, post & 6-month follow-up data)	N = 40 (N in each condition unclear). British study. Predominantly parents Patients pre-dominantly male, mean age = 36, 8 yrs since onset, 3.7 prior hospitalizations, all with schizophrenia.
Solomon et al. (1997 & 1996)	a) Ten-week group education or b) individual family consultation (at least 6 hrs)	Experimental: (2 intervention & 1 control conditions with pre, post & 9-month follow-up data)	N = 183 (group education = 47; individual consultation = 56; control = 80). Predominantly female & white family members, some college education, middle class; 76% parents. Patient mean age = 36, 13 yrs since onset, median no. prior hospitalizations = 4, majority with schizophrenia, some affective disorders.

Table 3. Psychoeducational Interventions with Families and Patients Interventions, Research Designs, and Subjects

Study	Interventions	Research Design*	Subjects (N = Individuals, not Families)
Anderson et al. (1986)	Three 4-hr meetings of a) professionally led support group or b) psychoeducational group (education, advice & coping skills)	Experimental: (2 conditions with pre & post data); plus a post-hoc no-treatment comparison group	N = 40 patients (20 per condition), all with affective disorders; primarily female, mean age = 50. Families mostly spouses, some parents.
Bentley (1990)	10 in-home sessions over 5 weeks (education plus communication & problem-solving skill training)	Four single-system designs (longitudinal data & pre-post data)	N = 4 patients with schizophrenia, ages 21-38, "good" premorbid functioning, 2-7 prior hospitalizations; 3 black; 3 living with family. Families: 3 mothers, 1 sister, ages 37-75.
Falloon et al. (1984, 1985 a & b); Falloon & Pederson (1985); Falloon (1986); Doane et al. (1986); McGill et al. (1983)	9-month in-home behavioral program of education (2 sessions) & structured problem-solving & communication skill training; optional monthly support group & post-9 mo. consulting sessions.	Experimental: (intervention & control conditions with pre, post, 3 & 9 mo. & 2-year follow-up data)	N = 36 patients with schizophrenia; predominantly male, mean age = 26, 4 yrs since onset, 3 prior hospitalizations; residence predominantly with family. Families primarily mothers. Ethnicity: 42% white, 36% black, 17% hispanic, 6% Asian. Middle to low socioeconomic class.
Glick et al. (1985, 1990, 1991); Clarkin et al. (1990); Haas et al. (1988; 1990); Spencer et al. (1988)	Weekly or semi-weekly psychoeducation & problem-solving individual family sessions (45-60 min. each).	Experimental: (intervention & control conditions with pre, post, 6 & 18-month follow-up data)	N = 169 inpatients. Diagnoses: schizo-phrenia (n = 92), affective disorders (n = 50), all other Axis I diagnoses (n = 27); mean age = 26; predominantly white, some black; mean inpatient stay = 51 days. No data on family characteristics.
Liberman et al (1984)	9-week behavioral training & educational multi-family groups (2 hrs each), focusing on problem-solving & communication skills.	Quasi-experimental: (intervention condition & 2 matched com-parison conditions: insight family therapy & standard treatment)	N = 14 patients with schizophrenia; all white males, mean age = 26, 3 prior hospitalizations, 6 yrs since onset, living with high EE family for at least 1 of 3 months prior to hospitalization. No data on family characteristics.

(continued)

Table 3 (Continued)

Study	Interventions	Research Design*	Subjects (N = Individuals, not Families)
Mills & Hansen (1991)	5-week groups, 3 hr/wk, in residential facility: a) psychoeducational groups or b) support groups, both professionally led.	Experimental: (2 intervention conditions with pre, post & 3-mo. follow-up data; Note: Ss/conditions not equivalent at baseline)	N = 12 patients with schizophrenia in residential facility, predominantly male, all white, 19-40 yrs old. Families all parents, 40-70 yrs old.
Mills et al. (1988)	Four 4-hr. bi-weekly groups focusing on education, communication & problem-solving skills	Quasi-experimental: (one condition with pre & post data)	N = 5 patients with schizophrenia, 4 males, living in residential facility. Four families were parental couples.
Mills et al. (1986)	10-week groups, 2 hrs/wk, in residential facility, focus on communication & conflict resolution (no education on schizophrenia & treatment mentioned).	Quasi-experimental: (one condition with pre, post & 3-month follow-up data)	N = 4 patients with schizophrenia, age 25-32, living in residential facility. All families were parental couples.
Zastowny et al. (1992)	16-week, 1.5 hr/wk, single family sessions of a) behavioral family management (see Falloon) or b) supportive family management (education, resource linkage, direct advice, support)	Experimental: (2 intervention conditions with pre, post & 6 to 16-month follow-up)	N = 30 patients in residential facility with schizophrenia, "high-risk chronic," mean age = 24, 3.7 prior hospitalizations, 4 yrs since onset, predominantly male. Predominantly white, middle class, two parent families.

Note: * All experimental designs used random assignment to conditions.

98

Table 3 presents the methodology of the nine psychoeducational interventions. Note that a number of the major psychoeducational intervention studies (Leff et al. 1982, 1985, 1989, 1990; Tarrier et al. 1988, 1989; McFarlane et al. 1993, 1995; Hogarty et al. 1986, 1991) are not included here because their reports did not include any results on family outcomes, the focus of this review. The psychoeducational interventions reviewed here almost always include an educational component (Mills et al., 1986, is the only exception to this), but all go beyond education to providing families with advice and/or training in communication and problem-solving skills. Variations reflect the different philosophical bases of the investigators. Although all subscribe to the stress-diathesis model of mental illness, many focused specifically on reducing expressed emotion (EE) in the family through communication and problem-solving training. Other interventions focus on providing families with direct advice about managing the illness and with local resources linkages, rather than training in specific skills. Many of the interventions attempt to provide social and emotional support for families' efforts to cope more successfully with the illness. Some of this support is implicit in the use of groups for the delivery of the intervention. Five of the interventions were provided in the context of multi-family groups.

As Table 3 indicates, the length of the psychoeducational interventions varied widely, however, the typical psychoeducational intervention is 9 to 10 weeks, exposing families to between 16 and 24 hours of intervention. All included both the mentally ill family member and other family members. Two were provided in the family home and the rest were provided in hospitals or residential facilities in which the mentally ill family member was residing at the time the intervention began.

Table 4 presents data on the individual family support and multi-component interventions and, as such, the interventions varied dramatically. Of the seven studies included in this category, five provided individualized assistance to families by professionals. This assistance was variously described as family consultation, individual counseling, or family support and education. Only brief descriptions of the interventions were provided in these reports making it difficult to gauge differences in types and amounts of assistance received between the five programs. The remaining studies in this category are two state-wide multi-component family support and education initiatives in Massachusetts and New Jersey. The Massachusetts study includes 12 programs offering from two to 10 different types of family support services, while the New Jersey study includes eight family support programs offering six different types of services. Services included individual family support, support groups, education, psychoeducation, as well as respite care, advocacy, patient companion programs, and technical assistance to family self-help groups.

Research Designs

Tables 1 to 4 also identify the research designs used in these studies and allow comparisons across categories. In particular, it can be seen in Tables 3 and 4 that studies in the two categories of Individual Family Support/Multi-Component Interventions and Psychoeducational Intervention studies utilized the strongest research designs, with over half of the studies (4 of 7 and 5 of 9 studies respectively) in these two categories utilizing experimental designs. Slightly over one-third (6 of 17 studies) of Educational Interventions utilized experimental designs, while only one-fifth (2 of 10) of Support Group studies used them. The quasi-experimental designs used across all four intervention modalities were fairly weak. The majority of them were a single treatment group with data collection at multiple time points, but no comparison group. While almost all studies collected pre-and post-intervention data, many did not collect any follow-up data. The strongest studies in terms of follow-up data were the Individual Family Support/Multi-Component Interventions with over two-thirds of the studies (5 of 7) in this category collecting follow-up data. Over half of the Educational and Psycho-educational Interventions (11 of 17 and 5 of 9 respectively) collected such data, but very few (1 of 10) of the Support Group studies did so. The lack of follow-up data is an important limitation of many of these studies, since previous research with a variety of interventions in the mental health field show that effects of some interventions tend to dissipate within 24 months after the end of the intervention period (Hogarty et al. 1991). Without follow-up data, practitioners and policy-makers have insufficient information with which to judge the persistence of treatment effect, particularly the need for further assistance, such as "booster sessions" that might benefit families and patients.

Overall, the weaknesses of most of the designs used in these studies limit our ability to conclusively attribute changes in outcome variables to the interventions examined. Nevertheless a sufficient number used rigorous designs and extensive follow-up periods to suggest potential benefits to families and their members with mental illnesses.

Subjects

It is difficult to draw overall conclusions about the subjects utilized in these intervention studies, since many provided incomplete data on the demographic characteristics of families and patients receiving these interventions. Based on the data presented, we can state that the typical family caregivers in most studies were white female parents of a male family member with schizophrenia, who were likely to be middle-class. However, there are clear exceptions to this. For example, most of the studies included fathers and other family members, even though they were rarely the majority. Many studies included patients with a wide variety of diagnoses, particularly the support group and educational interventions. In

Table 4. Individual Family Support and Multi-Component Interventions Interventions, Research Designs and Subjects

Study	Interventions	Research Design[*]	Subjects (N = Individuals, not Families)
Benson et al. (1996)	State-wide on-going program of 12 multi- component family support programs	Quasi-experimental: (intervention condition only with pre & post data at 6 mo.)	N = 647 families. Families: 75% Caucasian, more than half less than 20K annual income. Patients: most lived at home, diagnosed with schizophrenia or affective disorder.
Carpentier et al. (1992)	Year-long ongoing comprehensive psychosocial case management program for patients and families. Control condition: traditional case management program.	Quasi-experimental: (intervention condition & matched comparison condition with post only data; at least 12 mo. participation in either program).	N = 37 patients and their families (15 in comprehensive program and 22 in traditional case management program). Families: mean age =52 yrs., predominately female. Patients: mean age = 37, predominately male, diagnosis of schizophrenia.
Hoult et al. (1984); Reynolds & Hoult (1984)	a) community treatment (including family support and education) or b) hospital care and aftercare	Experimental: (2 intervention conditions with pre and post data at 4, 8 & 12 mo.)	N = 120 patients and families (intervention for patients; data collected from patients and families). Families: half parents, one-fifth non-relatives, majority lived with patient. Patients: majority female, less than 40 yrs. old, with schizophrenia
Riesser et al. (1991)	8 community programs offering six different family interventions (i.e., single family consultation, referral/linkage, psycho-educational group, support group, respite) for six mo.	Quasi-experimental: (intervention condition with pre & post at 3 and 6 months)	N = 191 families. Predominately female, parents, over 55 yrs, higher percentage in poverty than in state overall. Patient mean age = 36; predominately male, Caucasian, lived with family.

(continued)

101

Table 4 (Continued)

Study	Interventions	Research Design[*]	Subjects (N = Individuals, not Families)
Solomon et al. (1996, 1997)	a) single family consultation (6-15 hrs.) or b) education group (20 hrs. over ten weeks)	Experimental: (2 intervention conditions & control condition with pre, post at 3 & 6 mos.)	N = 183 (family consultation = 56, psycho-education = 47, control = 80). Families predominately Caucasian, female, parents, highly educated, mean age = 56 yrs. Patients: majority with schizophrenia.
Stein & Test (1980); Weisbrod, Test & Stein (1980); Test & Stein (1980)	Long-term assertive community treatment (incl.family support and education). Control condition: hospital care.	Experimental: (intervention and control conditions with pre & post at 1 & 4 mos.	At 1 mo., N = 49 families/patients (ACT = 28, control = 21); at 4 mo, N = 40 families (ACT = 22, control = 18). Patients: majority male, unmarried with a mean age of 30 yrs. No data on family characteristics.
Szmukler et al. (1996)	Individual counseling, 6 hrs. (one hr/wk.). Control group: 1 hr. information only group session	Experimental: (intervention and control conditions with pre and post at 3 & 6 mos.	N = 63 family caregivers; majority parents, 40+ yrs. Patients: majority lived with caregiver, mean age late 20s to early thirties.

Note: [*] All experimental designs used random assignment to conditions.

terms of the ethnicity of the samples, as can be seen in Table 2, a number of the Educational interventions were carried out by European investigators whose samples, we must assume, were fairly ethnically homogeneous reflecting the country of the investigators. Several American studies (e.g., Medvene et al. 1995; Norton et al. 1993; Falloon et al. 1984) had significant numbers of African-American and Hispanic families, increasing the generalizability of their results; however, they did not report results by ethnic group. The lack of diversity of subjects in many of these studies is an important issue given the large body of research documenting differences between groups of individuals in how they meet needs and solve problems. A variety of obstacles—systemic and personal—differentially affect the ability, opportunity, and willingness of specific population groups to participate in organized help-seeking and receiving activities (Giordano 1973; Naparstek, Biegel, and Spiro 1982; Neighbors and Jackson 1996).

OUTCOMES MEASURED

Tables 5 through 8 present the findings for each set of intervention studies, organized by type of outcome measured. The studies varied distinctively across intervention categories in the outcomes they measured which reflect their different goals. Given differences in outcomes, specific measures, and the language used to describe the outcomes the studies focused on, it is sometimes difficult to generalize across categories. The most typical type of outcomes assessed were variables that tapped indicators of family stress, its impact on the family and their efforts to deal with this situation. These variables include family burden, functioning, caregiver mental health status, and coping strategies. We will discuss each of these separately.

Family burden, distress or stress was measured by all four types of intervention studies. Specifically, all but one of the Individual Family Support/Multi-Component Interventions, less than half of the Educational interventions (8 of 17), and one-third of the Support Groups and Psychoeducational intervention studies (3 of 10 and 3 of 9, respectively) examined this variable. Family functioning, including family conflict (which we will distinguish from relationship with and attitude toward the patient, to be discussed shortly) was assessed by more than half of the Psychoeducational studies (5 of 9), half of the Support Group studies (5 of 10), one-third of the Education studies (6 of 17), and none of the Individual Family Support and Multi-Component interventions. Family members' mental health symptoms, including psychosomatic symptoms, were assessed by very few studies. Only three of the 10 Support Group studies, three of the 17 Educational interventions, one of the Psychoeducational, and two of the Individual Family Support and Multi-Component interventions measured family members' depression, anxiety or psychosomatic symptoms. This contrasts with the consistent finding from the caregiver burden literature that high

Table 5. Support Group Interventions for Families Outcomes Investigated by Two or More Studies

Study	Coping & Burden	Family Functioning and Environment	Caregiver & Patient Health and Mental Health	Caregiver Attitudes and Beliefs	Social Support/Social Network	Assessment of Support Group Experience
Anderson et al. (1986)	Improved coping with family member's illness reported	Decrease in dyadic consensus and affectional expression	No data	Positive changes in attitudes toward ill family member	No data	Members positive about group experience but wanted more information
Biegel & Yamatani (1986, 1987)	No data	No data	No data	No data	Most frequent help-giving activities pertained to non-threatening aspects of social support	High satisfaction with group, information and emotional support most helpful; sharing and interacting liked best; degree of group participation positively related to perceived helpfulness.
Heller et al. (1997)	No data	No data	No data	No data	Higher information benefit of group associated with low informal support and high levels of support provided and given during group meetings; greater relationship benefits of group associated with provision of support to other group members	Higher information benefit of group associated with low unmet service needs; greater relationship benefits of group associated with positive health

Study					
Kane et al. (1990)	No significant changes in levels of burden	No data	No significant changes in levels of caregiver mental health	No data	No significant changes in levels of social support	Only moderate levels of satisfaction with support group, fairly low ratings of satisfaction with information received
Medvene & Krauss (1989)	No data	Behavioral interactions with family members improved over time. Parental self-blame associated with less comfortable behavioral interactions	No data	Parents attributional processes influenced by group participation: organic attributions endorsed more strongly over time and psychogenic less.	No data	No data
Medvene et al. (1995)	Support group members had higher levels of future care burden and emotional health burden than non-group members	No significant differences in parent's causal attributions of mental illness between support group attenders and non-attenders	Support group attenders had family members with higher levels of symptomatology than non-attenders	No significant differences on causal attributions of mental illness between support group and non-group members	No significant differences in satisfaction with social support between support group attenders and non-attenders	No data
Mills & Hansen (1991)	No data	Support group participants had increased levels of family functioning	No data	No data	No data	No data

(continued)

Table 5 (Continued)

Study	Coping & Burden	Family Functioning and Environment	Caregiver & Patient Health and Mental Health	Caregiver Attitudes and Beliefs	Social Support/Social Network	Assessment of Support Group Experience
Monking (1994)	No data	No data	Support group participants had few physical complaints; more frequent attendees had family members with greater symptomatology and hospitalizations	No data	Support group members had higher levels of social contacts	No data
Norton et al. (1993)	No data	No data	No data	No data	No data	Support group members perceived greater personal benefits with no difference in the level of social/purposive benefits; social/organiz. Costs were seen as lower by support group members; group members reported a higher overall cost/benefit ratio
Potasznik & Nelson (1984)	Supportiveness of self-help group, total support satisfaction and density were negative and objective burden	Family conflict was positively associated with objective burden	No data	No data	Satisfaction with total social network was negatively correlated with network size and positively associated with supportiveness of the support group	See Burden & Social Support findings; Positive aspects of group included providing information, sharing of support and copying strategies, and a common bond among members (Q).

levels of burden are associated with such emotional sequelae (Biegel, Sales, and Schulz 1991), which suggests that reductions in depression, anxiety and physical symptoms should be targets of family interventions.

Conceptually speaking, family coping includes a wide range of strategies, some internal to the family such as family communication and problem-solving strategies, self-efficacy, and perceptions of ability to cope, and others more external such as acquiring knowledge about mental illness, obtaining and using social support, and using mental health services. A number of studies looked at internal strategies. This was most common among the Psychoeducational interventions (4 of 9 studies), the Individual Family Support and Multi-Component interventions (3 of 7), and the Educational studies (6 of 17). Only one of the 10 Support Group studies measured such variables.

Measures of knowledge of mental illness and its management was predictably most common in the Educational interventions. Almost all (15 of 17 studies) assessed knowledge. Only one-third (3 of 9) of the Psychoeducational interventions measured this variable and only one Support Group study did so. The Individual Family Support/Multi-Component Interventions did not measure this outcome at all.

Family caregivers' social support was only assessed by the Support Group interventions, although of the 10 studies, four did not measure this variable. None of the other three types of interventions measured this variable.

Use of mental health resources/services were assessed by three of the seventeen Educational interventions and three of the seven Individual Family Support/ Multi-Component Interventions (see Tables 6 & 8). Though there is considerable evidence that caregivers face barriers to use of services for their family members and themselves (Biegel, Johnsen, and Shafran 1997), clearly most studies were silent on the effect of these interventions on service use.

A second block of concepts assessed by these studies relate specifically to the patient, the family member with the mental illness. These are measures of patient functioning, particularly his/her symptoms, sometimes measured as relapse or hospitalizations, and measures of the family's relationship with the patient. Measures of patient functioning were characteristic of two-thirds of the Psychoeducational interventions (6 of 9 studies) and almost half the Educational interventions (7 of 17 studies). One Support Group study (Monking, 1994) assessed patient symptomatology and hospitalization, but none of the Individual Family Support/ Multi-Component Interventions did so. Measures of the family's attitude toward the patient (typically expressed emotion: criticism, hostility and/or overinvolvement) was measured by about half of the Psychoeducational and Educational interventions (5 of 9 and 9 of 17 studies respectively).

A majority of all studies across modalities reported results of families' satisfaction with the intervention they received. This was primarily a descriptive outcome pertaining to the degree to which subjects' enjoyed their intervention experience. As will be seen below, reports of subject satisfaction are not always accompanied

Table 6. Educational Interventions for Families Outcomes Investigated by Three or More Studies

Study	Family stress & Burden, Including Patient Symptoms	Beliefs about Family's Ability to Manage	Relationship with and Attitude Toward Patient	Knowledge of Schizophrenia	Utilization of Services	Satisfaction with Program
Abramowitz & Coursey (1989)	Significantly reduced trait anxiety, personal distress & family disruption, more active coping behaviors.	No change in perceived self-efficacy.	No change in frequency of negative feelings toward patient.	No data.	More community resources used at post-test by intervention group.	Most helpful aspects: contact with others, information & knowledge about resources.
Barrowclough et al. (1987)	No data.	No data.	No data.	Functional knowledge increased.	No data.	No data.
Berkowitz et al. (1984)	No data.	No data.	No data.	Increase over time; education condition had greater increase, maintained at follow-up.	No data.	No data.
Birchwood et al. (1992)	In all conditions, stress and fear reduced. At follow-up, only decrease in anxiety & fear maintained.	Increased optimism, maintained at follow-up.	No data.	In all conditions, increase over time; group education intervention had greatest increase; gains not maintained at followup. Homework had no effect.	No data.	No data.
Canive et al. (1996)	No decrease in subjective distress and burden.	No data.	No change in annoyance at patient's behavior or expectation of recovery.	Increase at post and follow-up.	No data.	No data.

108

Cazullo et al. (1989)	No data.	No data.	No data.	Increase over time.	No data.	No data.
Ehlert (1988)	In combination & problem-solving conditions, improved family climate; not maintained over time. No change in patient relapse or symptoms.	No data.	In combination & problem-solving conditions, reduction in rejecting attitudes & over-protection; not maintained over time.	In all conditions, increase over time; maintained at follow-up.	No data.	No data.
Glynn, Pugh & Rose (1993, 1990)	No data.	No data.	Better understanding of patient.	Increase over time; maintained at follow-up.	No difference in number of NAMI meetings attended.	Perceived support increased pre to post; dropped to pre-levels by follow-up.
Hugen (1993)	Lower family conflict; positive but no change in family self-blame; at 3 month follow-up, drop in hospitalization but no change in patient symptoms or social functioning.	No data.	Attitudes toward patient unchanged.	Increase over time; gains maintained.	Increase in patient participation in community services & activities.	High satisfaction levels, maintained at follow-up.
Kane et al. (1990)	Education condition caregivers less depressed. No difference between conditions and no change over time in social support, patient symptoms, and family distress.	No data.	No data.	Both groups increased; no difference between them.	No data	Education condition more satisfied.

(continued)

Table 6 (Continued)

Study	Family stress & Burden, Including Patient Symptoms	Beliefs about Family's Ability to Manage	Relationship with and Attitude Toward Patient	Knowledge of Schizophrenia	Utilization of Services	Satisfaction with Program
Kuipers et al. (1989; McCarthy et al., 1989)	No data on family distress or burden. No difference or change in patient symptoms or relapse; experimental condition patients had fewer disruptive behaviors.	No changes overtime or differences between conditions in perception of family's ability to cope with illness.	In education condition, drop in critical comments, greater intimacy & reciprocity.	No changes over time or differences between conditions.	No data.	Education condition supportive.
Mannion et al. (1994)	Improvements in personal distress.	No data.	Improvements in attitudes toward ill spouse.	Increase over time in knowledge of illness & of coping strategies.	No data.	No data.
Orhagen & d'Elia (1992 a & b)	Decrease in feelings of distress. No decrease in objective burden, including worries about suicide. No change in coping.	No data.	Decrease in criticism levels.	Increase over time.	No data.	Educational intervention very helpful, appreciated, would recommend to others.
Posner et al. (1992)	At follow-up, decrease in distress. No change over time or differences between conditions in family cohesion & a daptability, coping strategies. No difference in relapse.	No data.	No changes over time or differences between conditions in feelings toward patient.	Education condition showed increase, maintained over time.	No data.	Education condition more satisfied with patient care; difference not maintained at follow-up.

Sidley et al. (1992)	No change in family stress or burden; no change in patient symptoms.	Increased belief in importance of family role in treatment; no change in beliefs about patient control of symptoms or necessity of medications.	No data.	Increase in factual knowledge in both groups; no changes in functional knowledge.	No data.	No data.
Smith & Birchwood (1987)	Worry & fear reduced, but gains not maintained. Family stress reduced, but gains not maintained at follow-up. Family objective burden reduced from pre-test to follow-up. No change in patient symptoms.	No change over time or between groups.	No data.	For both conditions, increase over time; maintained at follow-up; education group acquired & maintained more knowledge than those receiving knowledge in mailed booklets.	No data.	No data.
Solomon et al. (1996, 1997)	No difference over time or between conditions in burden, social support, adaptive coping, and stress.	Self-efficacy increased in both conditions; education condition gains over controls not maintained.	No data.	No data.	No data.	No data.

by changes in family caregiver outcomes over time. Over two-thirds (5 of 7) of studies in the Individual Family Support/Multi-Component Intervention category examined this outcome, followed by almost two-thirds (6 of 10) of the Support Group Studies, over one-half (5 of 9) of Psychoeducational Interventions and less than half (7 of 17) of Educational Interventions.

RESULTS

The results of the studies reviewed are presented in Tables 5 through 8. It should be noted that our discussion of study results is obviously limited by whether the investigators chose to measure a particular outcome and whether follow-up data were collected. In other words, the lack of effect may be an artifact of studies not measuring or reporting a particular outcome, rather than not impacting on that outcome. We will begin with outcomes on burden, stress and distress. Of the four types of interventions investigated, only Support Groups showed no clear evidence of decreases in family burden. The other interventions—Education, Psychoeducation, and Individual Family Support/Multi-Component Interventions—all found reductions in family burden associated with the interventions. The lack of reduction of family burden found in the support group studies makes sense given previous research findings indicating that a measure of consumer behavioral problems was the strongest predictor of burden after controlling for other variables (Biegel, Milligan, Putnam, and Song 1994). Support groups do not have any formal component that addresses this issue and support group members report that the most frequent help-giving activities in their groups are related to nondirective, nonthreatening aspects of social support, and that the least frequent activities are those that were focused on behavioral change (Biegel and Yamatani 1987).

Among the studies of Educational interventions which obtained follow-up data on burden (10 of 17), burden reduction was found to be short-term. However, the effect of Psychoeducational interventions on family burden appears longer lasting, from the data gathered in their typically longer follow-up periods. Similarly, in the case of Individual Family Support/Multi-Component Interventions, reduction in family burden in some studies were reported at longer follow-up periods of six and 12 months.

On the whole, family functioning increased in the Psychoeducational and Educational interventions which measured this variable, but studies of Support Groups found no evidence of increased family functioning. Again, given the focus of support groups to provide information and emotional support to family caregivers, it is not surprising that support groups do not affect family functioning. They do not have this as a specific objective. The Psychoeducational studies primarily looked at family conflict, a surrogate indicator of family functioning, and found decreases in the experimental conditions and evidence that this effect persisted over time. Among the Educational groups the durability of this effect is unclear, as many of the studies did not utilize follow-up data.

It is more difficult to draw definitive conclusions about improvements in family caregiver's mental health status given the few studies in each modality which investigated this issue. Among the studies that did look at depression, anxiety and psychosomatic symptoms, both Support Groups and Educational interventions found improvements, as can be seen in Tables 5 and 6. The one Psychoeducational intervention which measured these variables (Falloon et al. 1985b) found decreases that persisted for two years. However, the Individual Family Support/ Multi-Component Interventions found no clear evidence of change.

Among those studies which investigated changes in internal family coping strategies, positive results are also found, although drawing conclusions must be tempered by the particular definition of coping strategies used in the individual studies. All modalities, except Individual Family Support/Multi-Component Interventions, found significant increases in family coping strategies, particularly communication, problem-solving skills and perceptions of self-efficacy. This appeared to be strongest among the Educational and Psychoeducational interventions.

Other coping strategies investigated were acquisition of knowledge, social support, and use of mental health resources. Knowledge of mental illness and its management increased in Educational, Psychoeducational and the one Support Group intervention which measured this. Again, this finding does not persist unless the intervention appears to be long-term or more intensive. Increases in social support were only measured and found among Support Group participants. The increased use of mental health resources were found among the Individual Family Support/Multi-Component Interventions and to some degree in the Educational interventions.

Decreases in patient symptomatology are associated with the Educational and Psychoeducational interventions (see Tables 6 and 7), the only studies which measured this outcome. Again these improvements do not necessarily appear to persist beyond the intervention, unless it is of long duration, that is, of nine months or more. Similarly, the family's attitudes toward and relationship with the patient improves in the Educational and Psychoeducational interventions and changes appear to persist with the lengthier psychoeducational interventions. Given the nature of serious mental illness and the complexities of interactions between families and their mentally ill members, it is not surprising that changes in consumer symptomatology and the family's relationship with the mentally ill member only persist in the lengthier interventions. As mentioned previously, families appear to be satisfied with all these interventions, appreciating the information, the support from professionals and other families, and advice on coping more successfully with this devastating chronic illness.

In summary, we reiterate that we are limited in our conclusions by the studies' choice of outcomes and designs. All four intervention modalities investigated changes in family burden and functioning. Here the evidence is strongest that

Table 7. Psychoeducational Interventions for Families and
Patients Outcomes Investigated by Three or More Studies

Study	Family Burden, Symptoms, Functioning	Family Conflict, Dissatisfaction	Family Coping Strategies	Attitude Toward Patient, Expressed Emotion	Knowledge of Mental Illness	Patient Symptoms & Functioning	Satisfaction & Evaluation of Program
Anderson et al. (1986)	No data.	Increase in patient's dyadic adjustment; decrease in spouse. No difference between conditions.	In both conditions, no change in coping strategies.	No data.	No data.	No data.	Participants very satisfied; support group wanted more information.
Bentley (1990)	No data.	2 of 4 had no significant change in family conflict; 1 increased; 1 decreased.	No data.	3 of 4 families decreased EE at post; 2 maintained gains at follow-up.	No data.	4 of 4 patients decreased symptoms at post; 2 of 4 maintained gains at follow-up. No change in social adjustment	No data.
Falloon et al. (1984, 1985a & b); Falloon & Pederson (1985); Falloon (1986); Doane et al. (1986); McGill et al. (1983)	At post & all follow-up points, intervention condition drop in family burden (objective & subjective), depression, anxiety & psycho-somatic symptoms.	No data.	At post for intervention condition, increase in use of problem-solving skills.	At post, intervention condition drop in negative emotional style.	In intervention condition, increase pre to post; gains maintained at 3, 9, & 24 mo. follow-up.	At post & all follow up, intervention patients had fewer symptoms & relapses and less severe & briefer exacerbations and fewer hospitalizations; social functioning improved.	In both conditions, almost all patients & families very satisfied with treatment.

Glick et al. (1985, 1990, 1991); Clarkin et al. (1990); Haas et al. (1988;1990); Spencer et al. (1988)	No effect at discharge; at 6 mo, families of female patients had decreased burden; at 6 & 18 mo, families of patients with major psychosis had decreased burden.	No data.	More positive attitudes toward patient at discharge, 6 mo. & 18 mo. follow-up for females with schizophrenia. More positive attitude at 18 mo. toward poorer functioning patients with schizophrenia.	No data.		Overall positive effect at discharge, 6 mo. & 18 mo. follow-up. Gender & diagnostic differences: at 18 mo. females better; males unaffected or worse, at discharge & 6 mo., improvement for those with affective disorders.	No data.
Liberman et al (1984)	No data.	Decrease in conflict in both conditions.	No data.	Criticism decreased in both family treatment conditions, greater decrease in psychoeducation condition.	Knowledge increased in psychoeducation condition.	Decrease in symptoms; 21% relapse at 9 mo. vs 56% in matched control group.	Almost all very satisfied.
Mills & Hansen (1991)	In both conditions, family functioning increased.	No data.	In both conditions, conflict resolution & communication skills increased.	No data.	No data.	Increase in acting independently & in perceived impact on family decision making.	No data.

(continued)

115

Table 7 (Continued)

Study	Family Burden, Symptoms, Functioning	Family Conflict, Dissatisfaction	Family Coping Strategies	Attitude Toward Patient, Expressed Emotion	Knowledge of Mental Illness	Patient Symptoms & Functioning	Satisfaction & Evaluation of Program
Mills et al. (1988)	No change in family functioning.	Decrease in conflict.	No data.	No data.	No data.	No data.	No data.
Mills et al. (1986)	Increase in family organization at post; at follow-up, increased communication.	No data.	No data.	No data.	No data.	No data.	3 of 4 found program useful.
Zastowny, et al. (1992)	In both conditions, family burden decreased.	In both conditions, family conflict decreased.	In both conditions, problem solving skills increased.	EE decreased in both conditions.	In both conditions, knowledge increased.	In both conditions, hospitalization decreased, symptoms and functioning improved significantly at post, some loss of gain at 16 mo.	Participants highly satisfied.

Educational and Psychoeducational interventions lead to a decrease in family burden and an increase in family functioning. Although the Individual Family Support/Multi-Component Interventions also found evidence of decreased burden, this is difficult to interpret given the diversity of the interventions in this grouping.

The evidence provided here does indicate that of these four modalities, Educational and Psychoeducational interventions appear to have the strongest impact on family burden, functioning, and coping strategies, including acquisition of knowledge (see Tables 6 and 7). Improvements in caregiver mental health status are also associated with Educational and Psychoeducational interventions. The more lengthy Psychoeducational interventions in particular are associated with decreases in patient symptomatology and improvements in the family's relationship with the patient. The successful outcomes of the Multi-Component Interventions (see Table 8) suggest the potential benefits to families of having choices in the services offered to them, particularly the option of respite care, the most frequently used service in both studies. Individual family consultation on the other hand has much more mixed results than any of the other modalities.

IMPLICATIONS FOR MENTAL HEALTH PRACTICE AND POLICY

The implications for practice of this review echo the work of Hatfield and others (Hatfield 1978; Lefley 1996) who have urged mental health practitioners to provide families with education, support, and concrete advice on managing the illness. Findings from a number of studies indicate that this is what families say they want but are not getting from mental health professionals (Holden and Lewine 1982).

Education is vital because there is much specific information that families need. The typical lay person knows little about the nature of mental illness, its etiology, medications, prognosis, and psychosocial management. Until an ill family member is diagnosed, the family is equally unaware of these issues. Bizarre behaviors are inexplicable and frightening. Lay persons obtain most of their ideas about what constitutes mental illness and what its course may be from the popular media. Without more comprehensive education, the family does not have one of the most critical tools for coping—knowledge. Without knowledge, families are more likely to be "stuck" in coping strategies that are not useful for themselves or their family member, whether those strategies are denial, blame, or inappropriate expectations of themselves and their family member. They have fewer tools for supporting their family member, lacking precise language for interacting with professionals, knowledge of the importance of medication, and understanding the role of mental health programs. The focus on specific advice to families that characterizes the Psychoeducational and Educational interventions, like knowledge about mental illness, also gives families more tools for managing the illness.

Table 8. Individual Family Support and Multi-Component Interventions for Families and Patients Outcomes Investigated by Two or More Studies

Study	Stress & Burden	Health and Mental Health Outcomes	Coping, Mastery & Self-Efficacy	Service Use	Relationship with Mental Health Professionals	Satisfaction with Intervention
Benson et al. (1996)	Stress reduced over time & associated with staff assessments of participation and hrs. of services used. All types of services used. Use of respite care only associated with strongest reduction in burden.	No data	No data	12 hrs. of service/month/ family across all pro-gramss. Two-thirds of all service hrs. used were respite.	Increased positive attitudes toward professionals.	Overall 94% of family members satisfied with programss; 97% would recommend to other family members.
Carpentier et al. (1992)	Lower burden in comprehensive program.	No data	No data	Receipt of more specific services associated with comprehensive program effects Greater need for services expressed by family members who didn't live with patient and by single parents.	More contact with mental health professionals in comprehensive program.	No data
Hoult et al. (1984); Reynolds & Hoult 1984)	Less worry about patient in community treatment condition; no differences between conditions in objective burden.	No data	Increased coping in community treatment condition.	No data	Higher satisfaction in community treatment condition with mental health professionals' advice, information, support, medication protocols, and supervision.	Higher satisfaction in community treatment group with care patient received. For families who lived with patients, community treatment more satisying & helpful than hospital care.

Riesser et al. (1994)	Reduction in overall burden at 6 months but not in areas most burdensome; small reduction in degree of concern for ill family member at 6 months. Reduced burden associated with greater number of different services used and with amount of family support and single family services used.	No data	No data	Most frequently used services: single family consultation & referral linkages. Increased involvement in family support groups; involvement associated with greater perceived support and less psychoeducation used. Greater involvement with NAMI also associated with increased family support and with decreased use of psychoeducation & systems advocacy.	No data	Significant increase in satisfaction with mental health services after six months. Greater satisfaction with services associated with amount of family support and single family services used.
Solomon et al. (1996, 1997)	No differences between conditions in stress.	No differences in levels of grief between conditions.	No change in individual family consultation condition in mastery; at 3 mo, increased self-efficacy, at 6 mo. no difference with control group	No data	No data	Families in individual consultation condition report intervention more helpful than did education group members in learning about community resources, but less helpful in understanding medication.

(continued)

Table 8 (Continued)

Study	Stress & Burden	Health and Mental Health Outcomes	Coping, Mastery & Self-Efficacy	Service Use	Relationship with Mental Health Professionals	Satisfaction with Intervention
Stein & Test (1980); Weisbrod, Test & Stein (1980); Test & Stein (1980)	No differences between conditions in decreased burden or family disruption. Within groups analysis shows that subjective burden decreased for Assertive Community Treatment condition.	No data	No data	No data	No data	No data
Szmukler et al. (1996)	No data	No differences between conditions or over time in general health, physical health, GP visits or affect.	No differences between conditions or over time in care-giving appraisal, coping or mastery.	No data	No data	More intervention condition participants said it helped them better understand the patient; 60% of all subjects would recommend the intervention to others.

Compared to individual family therapy approaches that avoid advice, these interventions have much greater impact because they use concrete knowledge from the empirical literature on mental health consumers' sensitivity to stress and identify the precipitants of relapse. Note that while the Support Group interventions did not examine changes in knowledge about mental illness, sharing information as well as emotional support is a frequently reported help-giving activity in self-help groups. Our knowledge of education and psychoeducational groups would be improved by documenting their impact on social support.

The isolation families of the mentally ill experience is well-documented (Doll 1976; Leff 1983). One of the great advantages of the group-based interventions (particularly the Support Groups and Educational interventions) is that families are exposed to others who are struggling with the same issues. From that common experience, their reactions are normalized and their efforts to cope encouraged. Although not documented quantitatively as the supportiveness of the Support Groups has been in their studies, informal and qualitative information from the studies of Educational interventions and group-based Psychoeducational interventions, suggest that families find these experiences supportive as well. Part of this appears to be due to the understanding and support of professionals, but a larger part is due to the exposure to other families of persons with mental illness.

The rationale for the focus in Psychoeducational interventions on changing family communication and problem-solving skills is still being debated by family advocates. Some argue that we have merely shifted from blaming families for causing their family member's mental illness to blaming them for causing a relapse (e.g., Solomon 1996). Professionals carrying out psychoeducational interventions need to be aware of families' sensitivity on this issue and frame their efforts in the knowledge that relapse is not infinitely preventable. The impact of psychoeducational interventions on patient relapse and symptoms is not arguable, given the large number of studies now which support this effect. And, as we have seen above, these interventions seem particularly effective in providing positive long-term effects for family caregivers as well.

Turning now to a discussion of the implications of these findings for mental health policy, we have seen that despite the limitations of the studies reported above, each of the four intervention modalities studied had some positive effects on families with some types of modalities having substantial, long-term positive effects. Unfortunately, despite these positive effects, interventions for family caregivers of persons with serious mental illness have not been widely adopted as standard practice by most mental health systems in the United States.

Family support groups are perhaps the most widely available intervention modality for families. Many have their origins in self-help efforts by families organized because of their dissatisfaction with mental health professionals and mental health systems. Over the years, an increasing number of mental health agencies have developed partnerships with families and have helped organize, sponsor and staff support groups. An examination of their membership, however, reveals that like many other self-help groups, they predominately serve a Caucasian, upper middle class population who are not representative of family caregivers overall. Even beyond this limitation, the data above suggests that support groups, by themselves, may be insufficient to address the needs of family caregivers. Rather, families need the options offered by a wide variety of intervention modalities including educational and psychoeducational interventions and individual family support.

The National Alliance for the Mentally Ill (NAMI) has been in the forefront of designing and offering educational programs for their members and for other families in the community. These efforts are naturally limited by resource constraints. Rather than duplicating these efforts, public mental health systems should work with local NAMI groups to expand these programs with specific attention to targeting the inclusion of underrepresented groups of family caregivers.

Psychoeducational interventions, as discussed above, have been tested in efficacy studies but not widely implemented in natural settings in the United States. They have been recommended as a "best practice" in the treatment of schizophrenia by the recent Schizophrenia PORT study (Lehman et al. 1998). However, despite this and evidence of the cost-effectiveness of McFarlane's multiple group psychoeducational intervention (McFarlane et al. 1995), McFarlane reports wider adoption of his model of multi-family psychoeducation outside the United States than in this country (McFarlane, personal communication). Mental health systems in the United States appear to be missing a significant intervention option that can clearly help patients and their families. Greater efforts need to be expended by public mental health systems to adopt these innovative interventions.

IMPLICATIONS FOR FUTURE RESEARCH

Clearly no one study can address all threats to internal and external validity, nor measure all variables associated with the weight of caring for a family member with mental illness. We do not mean to suggest that every study should do so. Nevertheless, the research reviewed here has a number of significant limitations which should be addressed by future efforts in this area. We specifically recommend that future studies should:

Utilize Clearly Articulated, Theoretically Based, Intervention Modalities. Whether one is comparing the results of multiple intervention studies, or desiring to replicate a particular intervention study, it is imperative that full and complete information be provided on the theoretical framework underlying the intervention as well as a description of the intervention components and issues addressed in implementing the intervention. When an intervention study is reporting the testing of an intervention that has been utilized in other settings or situations, measures of fidelity of the intervention should also be utilized and reported. Almost none of the reviewed studies provided any data whatsoever regarding the fidelity with which the intervention was carried out.

Utilize Stronger Research Designs. Wherever possible experimental designs should be utilized. Where this is unfeasible, comparison groups should be used with quasi-experimental designs. The lack of follow-up data in the majority of reviewed studies is a major limitation. Future studies should emphasize the collection of follow-up data at one or more points in time in order to assess the degree to which intervention effects are sustained over time. Such data can be extremely valuable in helping to determine the need for intervention "booster shots" or additional supports for families. In addition, studies that compare different intervention modalities utilizing the same sampling procedures and measurement indices are particularly desirable.

Examine Issues of Dosage and Timing of Interventions. The literature on interventions for family caregivers of persons with mental illness has insufficiently addressed issues of dosage and timing of interventions. For example, are there times in the caregiving career when particular interventions are more appropriate and effective than at other times? For interventions that are time-limited, how much of the intervention is needed over what periods of time?

Utilize a Broader Range of Intervention Modalities. There are a variety of interventions developed for family caregivers of persons with other chronic illnesses and their family members that have not yet been tested with families of persons with mental illness. In particular, technology-based interventions, such as telephone support groups and computer networking and bulletin boards have been developed for persons with cancer and AIDS and for caregivers of those with Alzheimer's disease (Galinsky, Schopler, and Abell 1997; Coon, Schulz, and Ory 1999). A high technology, home-based intervention might be particularly suitable to address barriers facing families in rural areas or those whose obligations or own disabilities restrict access to out-of-home interventions.

Utilize a Broader Range of Outcome Variables. As noted throughout the paper and as can be seen in Tables 5-8, while the range of individual and family level outcome variables examined by reviewed studies is broad, there are a number of gaps with some outcomes receiving much less examination than others in some intervention modalities. As mentioned previously, knowledge acquisition in support groups and social support in education and psychoeducation groups need to be investigated. In addition, many of the studies in the Support Groups and the Individual Family Support/Multi-Component Intervention categories utilized far fewer outcome measures than the other intervention types. Overall, there were a number of measures, in particular caregiver health and mental health status, burden, coping and social support, that were underreported and that should be included to a greater extent in future research studies. The lack of more extensive inclusion of these variables is puzzling given the extant knowledge concerning their importance. For example, research findings with a variety of chronic illnesses, including mental illness, indicate that family caregivers are at significant risk for depression (Song, Biegel, and Milligan 1997). Also, social isolation is known to be a common characteristic of families of persons with mental illness as well as a major predictor of caregiver well-being (Biegel et al. 1994; Leff 1983).

In addition, the outcome variables in the reviewed studies focused almost exclusively on effects of the intervention on the family caregiver and sometimes also the mentally ill family member. System level outcomes also need to be examined in future research. For example, very few of the reviewed studies carried out cost-benefit analyses, clearly an important issue given the growth of managed care and its associated cost constraints.

Utilize Larger, More Representative Samples. The subjects of family caregiver intervention studies should be more representative of the universe of family caregivers. When heterogeneous samples of caregivers are utilized, analyses should be made to test for differences by social class and ethnicity. In designing interventions and intervention studies, more attention needs to be paid to the facilitators and barriers that particular subgroups of family caregivers experience that affect their use or non-use of particular interventions.

Utilize a Combination of Quantitative and Qualitative Data. Wherever possible, standardized measurement instruments with established and acceptable levels of reliability and validity should be used. However, qualitative measures should also be included in studies as both an aid in interpreting the meaning of particular quantitative results as well as in examining variables which need to be measured in an exploratory fashion.

In order to fully address the above issues, large multi-site coordinated studies are needed. While the NIMH and the Center for Mental Health Services have funded a number of multi-site research studies in the mental health field, only one study included family interventions (Schooler, Keith, Severe, Matthews et al. 1997). However, there is a federal model for a family caregiver research program, the "Resources for Enhancing Alzheimer's Caregiver Health" (REACH) project. This unique five-year research program is sponsored by the National Institute of Aging and the National Institute for Nursing Research at the National Institutes of Health. Under this initiative, six research sites and a coordinating center have been funded to examine the effects of different types of theory-based caregiver interventions (Coon, Schulz, and Ory 1999). Utilizing a large and heterogeneous number of subjects in five different study sites, the REACH project is designed to test a variety of theory-based interventions, to specify the pathways through which the interventions produce outcomes, to develop a wide variety of standardized outcome measures that are used across a variety of intervention modalities, and to create a common database in order to compare and contrast the effectiveness of interventions across a variety of populations. Utilization of this type of large-scale study by the mental health field can directly address a number of the limitations noted above in the studies reviewed and could significance advance our knowledge about effective interventions for caregivers of family members with mental illness.

In closing, despite gaps in our knowledge base and the need for future research to address unanswered questions, the findings of our review nevertheless have documented a number of positive outcomes for specific family caregiver interventions. Policymakers and practitioners need to increase efforts to facilitate the more widespread adoption of family interventions with demonstrated effectiveness in order to better meet the needs of family caregivers of persons with mental illness, many of whom remain underserved and under involved by the mental health system.

ACKNOWLEDGMENTS

This research was supported in part through grants from the Cuyahoga County Community Mental Health Board and the Office of Program Evaluation and Research, Ohio Department of Mental Health.

NOTE

1. Medline, Sociofile, Psyclit, Social Work, CINAHL, ERIC, Social Science Index, Ageline.

REFERENCES

Abramowitz, I. A., and R. D. Coursey. 1989. "Impact of an Educational Support Group on Family Participants Who Take Care of Their Schizophrenic Relatives." *Journal of Consulting and Clinical Psychology* 57 (2): 232-236.

Anderson, C. M., S. Griffin, A. Rossi, I. Pagonis, D. P. Holder, and R. Treiber. 1986. "A Comparative Study of the Impact of Education versus Process Support Groups for Families of Patients with Affective Disorders." *Family Process* 25: 185-205.

Barrowclough, C., and N. Tarrier. 1984. "Psychosocial Interventions with Families and Their Effects on the Course of Schizophrenia: A Review." *Psychological Medicine* 14: 629-642.

Barrowclough, C., N. Tarrier, S. Watts, C. Vaughn, J. S. Bamrah, and H. L. Freeman. 1987. "Assessing the Functional Value of Relatives' Knowledge about Schizophrenia: A Preliminary Report." *British Journal of Psychiatry* 151: 1-8.

Benson, P. R., G. A. Fisher, A. Diana, L. Simon, G. Gamache, R. C. Tessler, and M. McDermeit. 1996. "A State Network of Family Support Services: The Massachusetts Family Support Demonstration Project." *Evaluation and Program Planning* 19 (1): 27-39.

Bentley, K. J. 1990. "An Evaluation of Family-based Intervention with Schizophrenia Using Single-system Research." *British Journal of Social Work* 20: 101-116.

Berkowitz, R., R. Eberlein-Fries, and L. L. J. Kuipers. 1984. "Educating Relatives about Schizophrenia." *Schizophrenia Bulletin* 10 (3): 418-429.

Biegel, D. E., J. A. Johnsen, and R. Shafran. 1997. "Overcoming Barriers Faced by African-American Families with a Family Member with Mental Illness." *Family Relations* 46 (2): 163-178.

Biegel, D. E., E. Milligan, P. Putnam, and L. Song. 1994. "Predictors of Burden among Lower Socio-economic Status Caregivers of Persons with Chronic Mental Illness." *Community Mental Health Journal* 30 (5): 473-494.

Biegel, D. E., E. Sales, and R. Schulz. 1991. *Family Caregiving in Chronic Illness: Alzheimer's Disease, Cancer, Heart Disease, Mental Illness, and Stroke*. Family Caregiver Applications Series, Vol. 1. Newbury Park, CA: Sage Publications.

Biegel, D. E., L. Song, and S. Milligan. 1995. "A Comparative Analysis of Family Caregivers' Perceived Relationships with Mental Health Professionals: A Comparative Analysis." *Psychiatric Services* 46 (5): 477-482.

Biegel, D. E., and H. Yamanti. 1986. "Self-help Groups for Families of the Mentally Ill: Research Perspectives." In *Family Involvement in the Treatment of Schizophrenia*, edited by M. Z. Goldstein. Washington, DC: American Psychiatric Press.

Biegel, D. E., and H. Yamatani. 1987. "Help-giving in Self-help Groups." *Hospital and Community Psychiatry* 38 (11): 1195-1197.

Birchwood, M., J. Smith, and R. Cochrane. 1992. "Specific and Non-specifc Effects of Educational Intervention for Families Living with Schizophrenia: A Comparison of Three Methods." *British Journal of Psychiatry* 160: 806-814.

Bruhn, J. G. 1977. "Effects of Chronic Illness on the Family." *The Journal of Family Practice* 4 (6): 1057-1060.

Canive, J. M., J. Sanz-Fuentenebro, C. Vazquez, C. Qualls, F. Fuentenbro, I. G. Perez, and V. B. Tuason. 1996. "Family Psychoeducational Support Groups in Spain: Parents' Distress and Burden at Nine-month Follow-up." *Annals of Clinical Psychiatry* 8 (2): 71-79.

Carpentier, N., A. Lesage, J. Goulet, P. LaLonde, and M. Renaud. 1992. "Burden of Care of Families Not Living with Young Schizophrenic Relatives." *Hospital and Community Psychiatry* 43 (1): 38-43.

Cazzullo, C. L., P. Bertrando, M. Clerici, C. Bressi, C. Da Ponte, and E. Albertini. 1989. "The Efficacy of an Information Group Intervention on Relatives of Schizophrenics." *The International Journal of Social Psychiatry* 35 (4): 313-323.

Clarkin, J. F., I. D. Glick, G. L. Haas, J. H. Spencer, A. B. Lewis, J. Peyser, N. DeMane, M. Good-Ellis, E. Harris, and V. Lestelle. 1990. "A Randomized Clinical Trail of Inpatient Family Intervention. V: Results for Affective Disorders." *Journal of Affective Disorders* 18: 17-28.

Coon, D., R. Schulz, and M. Ory. 1999. "Innovative Intervention Approaches for Alzheimer's Disease Caregivers." Pp. 294-324 in *Innovations in Practice and Service Delivery Across the Lifespan*, edited by D. Biegel and A. Blum. New York: Oxford University Press.

Davis, A., S. Dinitz, and B. Pasaminick. 1974. *Schizophrenics in the New Custodial Community*. Columbus: Ohio State University Press.

Dixon, L. B., and A. F. Lehman. 1995. "Family Interventions for Schizophrenia." *Schizophrenia Bulletin* 21 (4): 631-643.

Doane, J. A., M. J. Goldstein, D. J. Miklowitz, and I. R. H. Falloon. 1986. "The Impact of Individual and Family Treatment on the Affective Climate of Families of Schizophrenics." *British Journal of Psychiatry* 148: 279-287.

Doll, W. 1976. "Family Coping with the Mentally Ill: An Unanticipated Problem of Deinstitutionalization." *Hospital and Community Psychiatry* 27 (3): 183-185.

Ehlert, U. 1988. "Psychological Intervention for the Relatives of Schizophrenic Patients." Pp. 251-259 in *Advances in Theory and Practice in Behaviour Therapy*, edited by P. M. G. Emmelkamp, W. T. A. M. Everaed, F. Kraaimaat, and M. J. M. van Son. Amsterdam: Swets & Zeitlinger.

Falloon, I. R. H. 1986. "Behavioral Family Therapy for Schizophrenia: Clinical, Social, Family and Economic Benefits." In *Treatment of Schizophrenia: Family Assessment and Intervention*, edited by I. H. and K. H. E. M. J. Goldstein. Berlin: Springer Verlag.

Falloon, I. R. H., J. L. Boyd, and C. W. McGill. 1984. *Family Care of Schizophrenia: A Problem-solving Approach to the Treatment of Mental Illness*. New York: Guilford Press.

Falloon, I. R. H., J. L. Boyd, C. W. McGill, M. Williamson, J. Razani, H. B. Moss, A. M. Gilderman, and G. M. Simpson. 1985a. "Family Management in the Prevention of Morbidity of Schizophrenia." *Archives of General Psychiatry* 42: 887-896.

Falloon, I. R. H., J. L. Boyd, C. W. McGill, M. Williamson, , J. Razani, H. B. Moss, A. M. Gilderman, and G. M. Simpson 1985b. "Family Management in the Prevention of Morbidity of Schizophrenia: Clinical Outcomes of a Two-year Longitudinal Study." *Archives of General Psychiatry* 42: 887-896.

Falloon, I. R. H., and J. Pederson. 1985. "Family Management in the Prevention of Morbidity of Schizophrenia: The Adjustment of the Family Unit." *British Journal of Psychiatry* 147: 156-163.

Falloon, I., J. Boyd, H. Moss, C. Cardin, J. McGill, J. Razani, J. Pederson, and J. Doane. 1985. "Behavioural Family Therapy for Schizophrenia: A Controlled Two-year Study." Pp. 481-485 in *Psychiatry, the State of the Art. Vol. 7: Epidemiology and Community Psychiatry*, edited by R. B. R. W. and K. T. E. P. Pichot. New York: Plenum Press.

Galinsky, M. J., J. H. Schopler, and M. D. Abell. 1997. "Connecting Group Members through Telephone and Computer Groups." *Health and Social Work* 22 (3): 181-188.

Giordano, J. 1973. *Ethnicity and Mental Health*. New York: American Jewish Committee.

Glick, I. D., J. F. Clarkin, G. L. Haas, J. H. Spencer, and C. L. Chen. 1991. "A Randomized Clinical Trial of Inpatient Family Intervention. VI: Mediating Variables and Outcome." *Family Process* 30: 85-99.

Glick, I. D., J. F. Clarkin, J. H. Spencer, G. L. Haas, A. B. Lewis, J. Peyser, N. DeMane, M. Good-Ellis, E. Harris, and V. Lestelle. 1985. "A Controlled Evaluation of an Inpatient Family Intervention. I: Preliminary Results of the Six-month Follow-up." *Archives of General Psychiatry* 42: 882-886.

Glick, I. D., J. H. Spencer, J. F. Clarkin, G. L. Haas, A. B. Lewis, J. Peyser, N. DeMane, M. Good-Ellis, E. Harris, and V. Lestelle. 1990. "A Randomized Clinical Trial of Inpatient Family Intervention. IV: Followup Results for Subjects with Schizophrenia." *Schizophrenia Research* 3: 187-200.

Glynn, S. M., R. Pugh, and G. Rose. 1990. "Predictors of Relatives' Attendance at a State Hospital Workshop on Schizophrenia." *Hospital and Community Psychiatry* 41 (1): 67-70.

Glynn, S. M., R. Pugh, and G. Rose. 1993. "Benefits of a Relative's Attendance at a Workshop on Schizophrenia at a State Hospital." *Psychosocial Rehabilitation Journal* 16 (4): 95-101.

Grad, J. P. D., and P. M. D. Sainsbury. 1963. "Mental Illness and the Family." *Lancet* 1: 544-547.

Haas, G. L., I. D. Glick, J. F. Clarkin, J. H. Spencer, and A. B. Lewis. 1990. "Gender and Schizophrenia Outcomes: A Clinical Trial of an Inpatient Family Intervention." *Schizophrenia Bulletin* 16 (2): 277-292.

Haas, G. L., I. D. Glick, J. F. Clarkin, J. H. Spencer, A. B. Lewis, J. Peyser, N. DeMane, M. Good-Ellis, E. Harris, and V. Lestelle. 1988. "Inpatient Family Intervention: A Randomized Clinical Trial. II: Results at Hospital Discharge." *Archives of General Psychiatry* 45: 217-224.

Hatfield, A. B. 1978. "Psychological Costs of Schizophrenia to the Family." *Social Work*: 355-359.

Hatfield, A. B. 1986. "Semantic Barriers to Family and Professional Collaboration." *Schizophrenia Bulletin* 12 (3): 325-333.

Heller, T., J. A. Roccoforte, K. Hsiech, J. A. Cook, and S. A. Pickett. 1997. "Benefits of Support Groups for Families of Adults with Severe Mental Illness." *American Journal of Orthopsychiatry* 67 (2): 187-198.

Hogarty, G. E., C. M. Anderson, D. J. Reiss, S. J. Kornblith, D. P. Greenwald, R. F. Ulrich, and M. Carter. 1986. "Family Psychoeducation, Social Skills Training, and Maintenance Chemotherapy in the Aftercare Treatment of Schizophrenia. I: One-year Effects of a Controlled Study on Relapse and Expressed Emotion." *Archives of General Psychiatry* 43: 633-642.

Hogarty, G. E., C. M. Anderson, D. J. Reiss, S. J. Kornblith, D. P. Greenwald, R. F. Ulrich, and M. Carter. 1991. "Family Psychoeducation, Social Skills Training, and Maintenance Chemotherapy in the Aftercare Treatment of Schizophrenia. II: Two-year Effects of a Controlled Study on Relapse and Adjustment." *Archives of General Psychiatry* 48: 340-347.

Holden, D. F., and R. R. J. Lewine. 1982. "How Families Evaluate Mental Health Professionals, Resources and Effects of Illness." *Schizophrenia Bulletin* 8: 626-633.

Hoult, J., A. Rosen, and I. Reynolds. 1984. "Community Oriented Treatment Compared to Psychiatric Hospital Oriented Treatment." *Social Science and Medicine* 18 (11): 1005-1010.

Hugen, B. 1993. "The Effectiveness of a Psychoeducational Support Service to Families of Persons with a Chronic Mental Illness." *Research on Social Work Practice* 3 (2): 137-154.

Kane, C. F., E. DiMartino, and M. Jimenez. 1990. "A Comparison of Short-term Psychoeducational and Support Groups for Relatives Coping with Chronic Schizophrenia." *Archives of Psychiatric Nursing* 4(6): 343-353.

Kreisman, D., and V. D. Joy. 1974. "Family Response to the Mental Illness of a Relative: A Review of the Literature." *Schizophrenia Bulletin* 10: 34-57.

Kuipers, L., B. MacCarthy, J. Hurry, and R. Harper. 1989. "Counselling the Relatives of the Long-term Adult Mentally Ill. II: A Low-cost Supportive Model." *British Journal of Psychiatry* 154: 775-782.

Lam, D. H. 1991. "Psychosocial Family Intervention in Schizophrenia: A Review of Empirical Studies." *Psychological Medicine* 21 (2): 423-441.

Leff, J. P. 1983. "The Management of the Family of the Chronic Psychiatric Patient." Pp. 159-179 in *The Chronic Psychiatric Patient in the Community: Principles of Treatment*, edited b I. Barofsky and R. D. Budson. New York: SP Medical & Scientific Books.

Leff, J., R. Berkowitz, N. Shavit, A. Strachan, I. Glass, and C. Vaughn. 1989. "A Trial of Family Therapy vs. a Relatives Group for Schizophrenia." *British Journal of Psychiatry* 154: 58-66.

Leff, J., R. Berkowitz, N. Shavit, A. Strachan, I. Glass, and C. Vaughn. 1990. "A Trial of Family Therapy vs. a Relatives' Group for Schizophrenia: Two-year Follow-up." *British Journal of Psychiatry* 157: 571-577.

Leff, J., L. Kuipers, R. Berkowitz, R. Eberlein-Vries, and D. Sturgeon. 1982. "A Controlled Trial of Social Intervention in the Families of Schizophrenic Patients." *British Journal of Psychiatry* 141: 121-134.

Leff, J., L. Kuipers, R. Berkowitz, R. Eberlein-Vries, and D. Sturgeon. 1985. "A Controlled Trial of Social Intervention in the Families of Schizophrenic Patients: Two-year Follow-up." *British Journal of Psychiatry* 146: 594-600.

Leff, J. P., and C. Vaugh. 1985. *Expressed Emotion in Families: Its Significance for Mental Illness*. New York: Guilford Press.

Lefley, H. P. 1996. *Family Caregiving in Mental Illness*. Newbury Park, CA: Sage Publications.

Lehman, A. F., D. M. Steinwachs, L. B. Dixon, H. H. Goldman, F. Osher, L. Postrado, J. E. Scott, J. W. Thompson, M. Fahey, P. Fischer, J. A. Kasper, A. Lyles, E. A. Skinner, R. Buchanan, W. T. Carpenter, Jr., J. Levine, E. A. McGlynn, R. Rosenheck, and J. Zito. 1998. "Translating Research into Practice: The Schizophrenia Patient Outcomes Research Team (PORT) Treatment Recommendations." *Schizophrenia Bulletin* 24 (1): 1-10.

Liberman, R. P., I. R. H. Falloon, and R. A. Aitchison. 1984. "Multiple Family Therapy for Schizoprhrenia: A Behavioral, Problem-solving Approach." *Psychosocial Rehabilitation Journal* 7 (4): 60-77.

MacCarthy, B., L. Kuipers, J. Hurry, R. Harper, and A. LeSage. 1989. "Counselling the Relatives of the Long-term Adult Mentally Ill. I: Evaluation of the Impact on Relatives and Patients." *British Journal of Psychiatry* 154: 768-775.

Mannion, E., K. Mueser, and P. Solomon. 1994. "Designing Psychoeducational Services for Spouses of Persons with Serious Mental Illness." *Community Mental Health Journal* 30 (2): 177-190.

Mari, J. D. J., and D. L. Streiner. 1994. "An Overview of Family Interventions and Relapse on Schizophrenia: Meta-analysis of Research Findings." *Psychological Medicine* 24: 565-578.

McFarlane, W. R., E. Lukens, B. Link, and R. Dushay. 1995. "Multiple-family Groups and Psychoeducation in the Treatment of Schizophrenia." *Archives of General Psychiatry* 52 (8): 679-687.

McFarlane, W. R., E. Lukens, B. Link, R. Dushay, S. A. Deakins, E. J. Dunne, B. Horen, M. Newman, and J. Toran. 1993. "Psychoeducational Multi-family Groups: Are They a Treatment of Choice for Schizophrenia?" Presented at June Annual Forum, Center for Practice Innovations, Mandel School of Applied Social Sciences, Case Western Reserve University, Cleveland, Ohio.

McGill, C. W., I. R. H. Falloon, J. L. Boyd, and C. Wood-Siverio. 1983. "Family Educational Intervention in the Treatment of Schizophrenia." *Hospital and Community Psychiatry* 34(10): 934-938.

Medvene, L. J., and D. H. Krauss. 1989. "Causal Attributions and Parent-child Relationships in a Self-help Group for Families of the Mentally Ill." *Journal of Applied Social Psychology* 19 (17): 1413-1430.

Medvene, L. J., R. Mendoza, K. Lin, N. Harris, and M. Miller. 1995. "Increasing Mexican-American Attendance of Support Groups for Parents of the Mentally Ill: Organizational and Psychological Factors." *Journal of Community Psychology* 23 (4): 307-325.

Mills, P. D., and J. C. Hansen. 1991. "Short-term Group Interventions for Mentally Ill Young Adults Living in a Community Residence and Their Families." *Hospital and Community Psychiatry* 42 (11): 1144-1150.

Mills, P. D., J. C. Hansen, and B. B. Malakie. 1986. "Psychoeducational Family Treatment for Young Adult Chronically Disturbed Clients." *Family Therapy* 13 (3): 275-285.

Mills, P. D., J. C. Hansen, and J. Sbarbati. 1988. "Family Psychoeducational Treatment for Young Adult Clients." *Family Therapy* 15 (1): 59-68.

Monking, H. S. 1994. "Self-help Groups for Families of Schizophrenic Patients: Formation, Development and Therapeutic Impact." *Social Psychiatry and Psychiatric Epidemiology* 29: 149-154.

Naparstek, A., D. Biegel, and H. Spiro. 1982. *Neighborhood Networks for Humane Mental Health Care.* New York: Plenum Press.

Neighbors, H. W., and J. S. Jackson (Eds.). 1996. *Mental Health in Black America.* Newbury Park, CA: Sage Publications.

Norton, S., A. Wandersman, and C. R. Goldman. 1993. "Perceived Costs and Benefits of Membership in a Self-help Group: Comparisons of Members and Non-members of the Alliance for the Mentally Ill." *Community Mental Health Journal* 29 (2): 143-160.

Orhagen, T., and G. d'Elia. 1992a. "Multifamily Educational Intervention in Schizophrenia. I: Does It Have Any Effect?" *Nord Journal of Psychiatry* 46 (1): 3-12.

Orhagen, T., and G. D'Elia. 1992b. "Multifamily Educational Intervention in Schizophrenia. II: Illness-related Knowledge and Mutual Support in Relation to Outcome." *Nord Journal of Psychiatry* 46 (1): 13-18.

Penn, D. L., and K. T. Mueser. 1996. "Research Update on the Psychosocial Treatment of Schizophrenia." *American Journal of Psychiatry* 153 (5): 607-617.

Posner, C. M., K. G. Wilson, M. J. Kral, S. Lander, and R. D. McIlwraith. 1992. "Family Psychoeducational Support Groups in Schizophrenia." *American Journal of Orthopsychiatry* 62 (2): 206-218.

Potasznik, H., and G. Nelson. 1984. "Stress and Social Support: The Burden Experienced by the Family of a Mentally Ill Person." *American Journal of Community Psychology* 12 (5): 589-607.

Reynolds, I., and J. E. Hoult. 1984. "The Relatives of the Mentally Ill: A Comparative Trial of Community-oriented and Hospital-oriented Psychiatric Care." *Journal of Nervous and Mental Disease* 172 (8): 480-489.

Riesser, G., S. Minsky, and B. Schorske. 1991. *Intensive Family Support Services: A Cooperative Evaluation.* New Jersey: Department of Human Services, Division of Mental Health and Hospitals, Bureau of Research and Evaluation.

Schooler, N. R., S. J. Keith, J. B. Severe, and S. M. Matthews. 1995. "Maintenance Treatment of Schizophrenia: A Review of Dose Reduction and Family Treatment Strategies." *Psychiatric Quarterly* 66 (4): 279-292.

Schooler, N. R., S. J. Keith, J. B. Severe, S. M. Matthews et al. 1997. "Relapse and Rehospitalization During Maintenance Treatment of Schizophrenia: The Effects of Dose Reduction and Family Treatment." *Archives of General Psychiatry* 54 (5): 453-463.

Sidley, G. L., J. Smith, and K. Howells. 1991. "Is It Ever Too Late to Learn? Information Provision to Relatives of Long-term Schizophrenia Sufferers." *Behavioral Psychotherapy* 19: 305-320.

Smith, J. V., and M. J. Birchwood. 1987. "Specific and Non-specific Effects of Educational Intervention with Families Living with a Schizophrenic Relative." *British Journal of Psychiatry* 150: 645-652.

Solomon, P. 1996. "Moving from Psychoeducation to Family Education for Families of Adults with Serious Mental Illness." *Psychiatric Services* 47 (12): 1364-1370.

Solomon, P., J. Draine, E. Mannion, and M. Meisel. 1996. "Two Modes of Brief Family Psychoeducation and Their Impact on Self-efficacy Regarding a Mentally Ill Relative." *Schizophrenia Bulletin*.

Solomon, P., J. Draine, E. Mannion, and M. Meisel. 1997. "Effectiveness of Two Models of Brief Family Education: Retention of Gains of Family Members of Adults with Serious Mental Illness." *American Journal of Orthopsychiatry* 67 (2): 177-186.

Song, L., D. E. Biegel, and E. Milligan. 1997. "Predictors of Depressive Symptomatology among Lower Social Class Caregivers of Persons with Chronic Mental Illness." *Community Mental Health Journal* 33 (4): 269-286.

Spencer, J. H., I. D. Glick, G. L. Haas, J. F. Clarkin, A. B. Lewis, J. Peyser, N. DeMane, M. Good-Ellis, E. Harris, and V. Lestelle. 1988. "A Randomized Clinical Trial of Inpatient Family Intervention. III: Effects at 6-Month and 18-Month Follow-ups. *American Journal of Psychiatry* 145 (9): 1115-1121.

Stein, L. I., and M. A. Test. 1980. "Alternatives to Mental Hospital Treatment. I: Conceptual Model, Treatment Program, and Clinical Evaluation." *Archive of General Psychiatry* 37: 392-397.

Szmukler, G. I., H. Herrman, S. Colusa, and A. Benson. 1996. "A Controlled Trial of a Counseling Intervention for Caregivers of Relatives with Schizophrenia." *Social Psychiatry and Psychiatric Epidemiology* 31(3-4): 149-155.

Tarrier, N., C. Barrowclough, C. Vaughn, J. S. Bamrah, K. Porceddu, S. Watts, and H. Freeman. 1988. "The Community Management of Schizophrenia: A Controlled Trial of a Behavioural Intervention with Families to Reduce Relapse." *British Journal of Psychiatry* 153: 532-542.

Tarrier, N., C. Barrowclough, C. Vaughn, J. S. Bamrah, K. Porceddu, S. Watts, and H. Freeman. 1989. "The Community Management of Schizophrenia: A Two-year Follow-up of a Behavioural Intervention with Families." *British Journal of Psychiatry* 154: 625-628.

Test, M. A., and L. I. Stein. 1980. "Alternative to Mental Hospital Treatment. III: Social Cost." *Archives of General Psychiatry* 37: 409-412.

Weisbrod, B. A., M. A. Test, and L. I. Stein. 1980. "Alternative to Mental Hospital Treatment. II: Economic Benefit-cost Analysis." *Archives of General Psychiatry* 37: 400-405.

Zastowny, T. R., A. F. Lehman, R. E. Cole, and C. Kane. 1992. "Family Management of Schizophrenia: A Comparison of Behavioral and Supportive Family Treatment." *Psychiatric Quarterly* 63 (2): 159-186.

THE CONTINGENT RATIONALITY
OF HOUSING PREFERENCES:
HOMELESS MENTALLY ILL PERSONS'
HOUSING CHOICES BEFORE AND
AFTER HOUSING EXPERIENCE

Russell K. Schutt and Stephen M. Goldfinger

ABSTRACT

Prior research indicates that homeless consumers of mental health services have a marked preference for independent living, while clinicians tend to recommend staffed, group housing. In order to understand this divergence and identify its consequences for mental health policy, we test influences on housing preferences suggested by rational choice and social structural perspectives. We use a randomized field trial of independent and group housing to identify the consequences for subsequent housing loss and consumer functioning of consumer- and clinician-determined housing placements. We find that consumer preference at baseline for independent living indicates greater vulnerability to housing loss, but the bases of preferences change after consumers gained experience with stable housing. We interpret the results as indicating the contingent rationality of preferences and the susceptibility of preferences to change with experience.

Research in Community and Mental Health, Volume 11, pages 131-156.
Copyright © 2000 by JAI Press Inc.
ISBN: 0-7623-0671-8

To provide mental health service consumers with the type of housing they want, rather than only what clinicians think they need (Drake and Adler 1984; Carling 1990; Anthony and Blanch 1989), has become a widely accepted goal for mental health service systems—albeit one that is still debated and is honored more often in theory than in practice (Minsky et al. 1995). In spite of a growing body of descriptive research, however, few scholars have attempted to explain variation in consumer housing preferences, to connect preferences research to broader theories of help seeking, or to test the relations between housing preferences and housing outcomes. As a result, the consequences of using preference-based decision making are not adequately understood and the potential for maximizing housing outcomes has probably not been achieved.

In this chapter, we develop and test a theoretically based explanation of housing preferences and we compare the consequences for consumer outcomes of preference-based and clinician-based placement. We first review descriptive research findings and theories of help seeking that can guide explanatory research on housing preferences. This work serves as background for our investigation of the housing preferences of homeless mentally ill persons who were recruited in homeless shelters and assigned randomly to either group homes or independent apartments.

We measured our consumers' housing preferences prior to housing placement and then 6, 12, and 18 months later. In this analysis, we test influences on these preferences at both baseline and the 18-month followup. Because we used random assignment to either group or independent living, we are also able to identify the independent effects of preferences and housing type on housing retention. Because clinicians rated our consumers' readiness for independent living, we are able to contrast the consequences of basing housing placement decisions on consumers' preferences with those made solely by relying on clinicians.

PREFERRED AND RECOMMENDED HOUSING

Between 60 and 70 percent of mental health service consumers choose independent living when they are offered a choice between independent apartments and staffed group homes (Tanzman 1993). When preferences for specific residential features are measured, it appears that some form of staff support is accepted, even desired, by almost two-thirds of consumers, particularly if it does not involve live-in staff or high levels of "behavioral demands" (Keck 1990; Minsky et al. 1995; Owen et al. 1996; Schutt and Goldfinger 1996; Schutt et al. 1992; Tanzman 1993). It is the group living feature of group homes that most consumers seek to avoid (Goering et al. 1990). However, when Minsky and colleagues (1995) distinguished different types of roommates, they found that living with family members or roommates of one's choice were each preferred by about one-third of consumers, but only 8 percent wanted to live with other mental health consumers.

Although there is thus some variation in consumer housing preferences, in general they contrast markedly with clinicians' housing recommendations. When asked to choose, clinicians recommend some type of supported group living for between 60 percent and 80 percent of consumers (Goldfinger and Schutt 1996; Minsky et al. 1995). Moreover, there appears to be little correspondence between the type of housing that consumers prefer and what clinicians recommend for those same consumers (Goldfinger and Schutt 1996).

EXPLAINING HOUSING PREFERENCES

The housing preferences of seriously mentally ill persons can be conceptualized as one aspect of their orientation to help seeking. We focus on two general alternative theories of help seeking in order to identify possible explanations for variation in preferences: rational choice theory and social structural theories, variously termed structural (Burt 1982), social organizational (Pescosolido 1992), socioeconomic (Etzioni 1988), and natural (Fjellman 1976). In brief, rational choice theory predicts that the rational weighing of costs and benefits will guide individuals' decisions (Coleman 1990), while social structural theories emphasize the role of social networks, the influence of emotions, and the interdependence of different decisions (Pescosolido 1992).

Rational choice theory expects consumers to make decisions that reflect their calculation of what is in their own interests. This assumption is exemplified by the "Health Belief Model," which posits that the propensity to seek health care or follow medical advice is maximized by perceptions of susceptibility to a disease and its likely severity, and by the anticipated benefits of and barriers to taking action. A more recent version, the "Health Decision Model," adds influences on health beliefs to its model of health decisions. In the Health Decision Model, background factors such as education and health insurance, and current social interaction shape experience with and knowledge about health. Experience and knowledge are both expected to influence general health beliefs and health care preferences (Eraker et al. 1984). Like Becker's (1996) extension of the microeconomic theory of preferences to take account of past experiences and social relations, the Health Decision Model recognizes the importance of social factors without altering the basic assumption that decisions are made on rational grounds.

Research in the general population has identified some of the influences predicted by the Health Decision Model. Use of mental health services is higher among those who report physical and mental health problems. The propensity to seek help for health and substance abuse problems also increases with the perceived severity of the disorder and its adverse social and health consequences (Hajema et al. 1999; Longshore et al. 1997). These influences interact with others: when severity is high, individuals are more likely to seek help irrespective of their other characteristics or network involvements (Frank and Kamlet 1989).

Background factors are also influential. Most studies of psychiatric help-seeking find that likelihood of seeking help for psychiatric problems increases with education (Kulka et al. 1979), perhaps because of a concomitant increase in an introspective, individualistic "psychological orientation" (Greenley and Mechanic 1976). In addition, the propensity to seek help from professionals increases with age and is higher among women than among men (Frank and Kamlet 1989; McKinlay 1972).

Pescosolido's (1992) "social organization strategy" framework exemplifies the developmental, contextual, and social assumptions of the social structural approach. It assumes that health care decisions must be understood as an ongoing series rather than as discrete events. It broadens the focus from decisions about formal medical care to the entire range of options that consumers consider for responding to health problems. It expects that individuals make health care decisions in interaction with others, and that they do so within the context of a particular time and place (Stoner 1985).

As predicted by social structural theory, empirical research has identified social network influences on help seeking. Individuals who are members of groups supportive of psychotherapy or who have friends with experience in psychotherapy are more likely to seek professional help for perceived mental disorders (Horwitz 1977; Kadushin 1969). However, individuals who are in cohesive informal networks, such as those associated with closely knit kin relations in stable communities, are less likely to seek professional services (Horwitz 1977; McKinlay 1972; Raphael 1964). Independent of the informal social network, positive contacts with nurses, social workers or other health system representatives may increase the propensity to seek professional help (Raphael 1964), while unpleasant contacts may lessen the likelihood of additional help-seeking (Breakey 1987; Kellerman et al. 1985).

In the case of housing preferences, there is some indication that the desire for autonomous living is associated with impairments that are likely to interfere with client autonomy (Morse et al. 1994), rather than, as predicted by rational choice theory, by ability to manage on one's own. Substance abuse predicted desire for independent living in our sample of homeless mentally ill persons, although so did perceived ability to manage on one's own (Schutt and Goldfinger 1996). Lovell and Cohen (1998) reported that some homeless consumers used their new housing as a source of goods to be resold to support a drug habit, rather than as a means for becoming residentially stable. In addition, some evidence suggests that clinician housing recommendations, which are not correlated with consumer preferences, reflect at least in part an assessment of the extent to which the consumer is capable of independent living. Dixon and colleagues (1994) found that psychotic symptoms and a diagnosis of schizophrenia were associated with clinicians' judgments that a client was unsuited for a Section 8 independent housing certificate. Consumers in our study were more likely to lose housing during the

18-month followup if at baseline clinicians had judged them to be not ready for independent living (Goldfinger et al. 1999).

AN INTEGRATED APPROACH

Although they present alternative perspectives on preferences, the predictions of rational choice and social structural theories are not mutually exclusive. The extension of rational choice theory by Becker (1996) and others brings social factors into the prediction of preferences to a limited extent—as an "overlay" to individual calculations of self-interest (Pescosolido 1992). Social structural explanations do not neglect the influence of rational calculation; rather, they note the limitations on individual rationality and consider additional influences on preferences.

We add to this nascent integration of rational choice and social structural theories the additional proposition of contingent rationality. We expect that consumer ability to make housing or service decisions on a rational basis is contingent on knowledge of the available alternatives. Without adequate knowledge, decisions are more likely to reflect social influences such as emotions, social relations, and alternative experiences. As consumers become familiar with the meaning of different housing types, whether as a result of experience or instruction, they are able to base their choices on "situated understanding" rather than just on abstract models (Lovell and Cohen 1998).

We hypothesize that both consumer residential preferences and clinician housing recommendations will influence housing outcomes, but for different reasons and in somewhat different ways at baseline and followup.

Specifically, at baseline we hypothesize that:

(1) Consumers whose housing placement is discrepant with their preference will have poorer housing outcomes. In other words, we expected our subjects to act like other consumers, more frequently leaving a housing option that they did not want.

(2) Consumers who receive independent housing but were recommended by clinicians for group housing would have poorer housing outcomes. We expect clinicians' recommendations to reflect perceived levels of consumers' need for support. Those consumers who did not receive needed supports would be less likely to maintain housing.

(3) Consumers who prefer independent housing but who were recommended by clinicians for group housing would experience poorer housing outcomes. These consumers, we felt, would be more resistant to services. We anticipate that impairments such as substance abuse and poor living skills would be more common in this subset.

We hypothesize that at followup:

(4) The housing preferences of those who remain in housing of a given type will become more favorable toward housing of that type. This reflects the basic expectation of the social structural model that social experience will influence preferences.

(5) Housing preferences at followup are more likely than at baseline to reflect consumer recognition of constraints on ability to live independently. After experience in housing, we expect that consumers will better recognize how well they can cope on their own with the demands of independent living.

METHODS

Project Description

The Boston McKinney Project developed two distinct residential alternatives: Independent Living (IL) and Evolving Consumer Households (ECHs), recruited homeless mentally ill persons to live in these residences and followed them for a total of 18 months after random assignment to one of the two residential types. Subjects paid 30 percent of their income for rent and utilities at all placement sites. The housing types compared were both expected to have some appeal to homeless mentally ill persons. Both were designed to maximize housing retention, but they used different arrangements to do so. Independent Living (ILs) units were both scattered site apartments and single room units in buildings with some communal space. None offered any on-site programming or resident clinical staff.

Evolving Consumer Households (ECHs) began as traditional, fully staffed group homes. Each accommodated six to 10 subjects, with virtually all in separate bedrooms, but with shared living, dining, recreational and kitchen facilities. Over time, tenants were encouraged by a residential coordinator to assume greater control of the residence. Clients were to help set household routines and policies and gradually take over many of the staff functions, from household chores to helping to manage the household, to making all decisions as an autonomous group and reducing the number of staff.

Study Design

Subjects were drawn from homeless individuals with serious and persistent mental illnesses between the ages of 18 and 65, living in transitional shelters operated by the Massachusetts Department of Mental Health. All English-speaking residents of the transitional shelters who passed a safety screen were asked to participate in the project. Project staff interviewed clients who consented to participate with an initial assessment schedule. Interview results were reviewed with the Project Director and Principal Investigator to ensure that all study subjects were homeless, seriously mentally ill, and had given informed and voluntary consent. Potential subjects judged to be at imminent risk of danger to themselves or others if placed in an independent apartment were excluded and reconsidered after six months (Goldfinger et al. 1996).

Eligible subjects entered a period of data collection, education, and training prior to housing placement, including a two week long standardized Housing Education Forum. Client and clinician housing preference questionnaires were also completed prior to housing placement. Follow-up interviewing and testing were done every six months for the next 18 months and data on residential status were collected throughout the project.

All subjects were provided an intensive clinical case manager (ICCM) with whom they met regularly over several weeks to prepare for housing transition. These case managers continued to assist subjects throughout the project. Study participants were not required to participate in psychiatric treatment, although they were encouraged by their case managers to use clinical services at their community mental health center (CMHC) that were appropriate to their needs and level of functioning. Through their CMHC, each subject had available to them an array of treatments—medication clinics, individual and group therapies, day programs for rehabilitation and social day activities, and, if needed, crisis intervention and hospitalization at the CMHC.

The Sample

A total of 303 transitional shelter residents were screened. Fifteen individuals were excluded because they were not able to speak and understand English or because they were not mentally ill (Goldfinger et al. 1996a). An additional 110 were deemed imminently dangerous and so were not recruited; of the remainder, 156 individuals provided informed consent to participate in the study. Ultimately 118 individuals were randomly assigned into ECH (*N*=63) or IL (*N*=55) sites. The remaining 38 either found alternative housing during the waiting period before placement or refused to move into the housing to which they were randomly assigned. Of the 118 subjects housed, a total of six were excluded from all longitudinal analyses (3 died, 1 had left the state prior to data collection, 2 were not allowed to move into the randomly assigned IL due to laws excluding convicted felons from publicly funded housing). Among the 118 housed study participants, 41 percent were African American, 72 percent were male, two-thirds had completed high school or a GED, and their average age was 38. Researchers using the Structured Interview for DSM-III-R identified schizophrenia or schizoaffective disorder as the primary diagnosis for 62 percent of the study participants. About half were identified as substance abusers.

Measures

Respondents answered ten questions concerning their residential preferences, with specific questions focusing on interest in moving, interest in staff support, and interest in living with others (Schutt and Goldfinger 1996). We then averaged the responses to nine of these questions to create a preference index, which had a Cronbach's alpha of .72 (one of the ten preference questions was not correlated with the others, and so was not included in the index).

Two specific index questions are used for analyses of the preference components. To assess desire for staff assistance, consumers were asked: How would you feel about having someone to help you with the things you have a hard time managing alone?Responses ranged from "Like the idea a lot" [1] to "Dislike the idea a lot" [5].

To assess desire for living with others as in the ECH, consumers were asked: If you had a choice of living with six or seven other people, where you had your own bedroom, or living in a small two-room apartment, which would you prefer? "Group living" [1] "Apartment" [2].

A systematic and realistic process was used to measure clinician residential recommendations for the consumers' housing placements (Goldfinger and Schutt 1996). Two experienced clinicians (a psychiatric nurse and a psychiatric social worker) who did not know the study subjects interviewed these consumers and reviewed their clinical records before the consumers were assigned to project housing. The clinicians answered eight questions regarding the desirability of particular housing features for each consumer and also stated whether, overall, they thought ECHs (group) or apartments (independent living) would be preferable. We averaged responses by the two clinicians to these eight questions (the average correlation was .43) and then computed our index of clinician residential recommendations as the average score across the eight items. Cronbach's alpha was .84.

Ability to manage tasks associated with independent living was measured with responses to nine questions about "things that people may have to do when they live in their own place," such as shopping, cleaning, and cooking. Cronbach's alpha was .69.

Psychiatric diagnosis was determined with the Structured Clinical Interview for DSM-III-R-Patient Edition (SCID-P, 9/1/89 Version (Spitzer et al. 1989)) and then recoded to a dichotomy that distinguished schizophrenia and schizoaffective disorders from other diagnoses (which were almost all major affective disorders). The SCID-P interviews were conducted by doctoral-level clinical psychologists, for whom inter-rater reliability was established at 100 percent for a subset of seven cases. We also used the SCID-P to identify lifetime substance abuse. Drug and alcohol abuse during the project was operationalized as the mean from the three follow-ups of a composite index based on case manager reports and self-reports, each scored to distinguish abuse, some use, and nonuse (Goldfinger et al. 1996b).

Housing status was recorded using client self-report, project housing and DMH records, and weekly case-manager logs. From these sources, a housing timeline was constructed that indicated the time spent by each subject in project housing, in other community housing, shelters, on the streets or in institutional settings such as hospitals, detox centers or jails. These data were compiled even for subjects who left project-sponsored housing or withdrew entirely from active participation in follow-up research. We calculated from this time the proportion of days spent homeless out of all days not institutionalized.

Every six months, ICCMs also completed the Life Skills Profile (Rosen et al. 1989) for each subject, based on their observations of the subject's behavior. We averaged the scale scores for each administration to generate average scores for each of the Life Skills dimensions: nonturbulence (Cronbach's α = .85), self-care (α = .78), social contact (α = .82), communication (α = .53), and responsibility (α = .75).

Other measures used in the analysis are a self-report inventory of psychiatric symptoms (the Colorado Symptomology Index) (Shern et al. 1994). Inter-item reliability for the index (Cronbach's alpha) was a very respectable .88. Subjective feelings of social support were measured with the Interpersonal Support Evaluation List (alpha = .90) (Cohen et al. 1985). The Arizona Social Support Inventory (Barrera 1981) was used to identify the number of positive and negative social relations.

Ninety-eight of the 118 subjects who moved into housing completed the Minnesota Multiphasic Personality Inventory 2. Personality characteristics were assessed with factor scores produced by a principal components analysis of the MMPI-2's 10 clinical scales (Greene 1991): psychoticism (most correlated with Mania, Schizophrenia, Paranoia, and Psychasthenia), anxiousness (Hypochondriasis, Hysteria, Depression), and withdrawal (Social Introversion, Depression). These three factors accounted for 74 percent of the common variance in the clinical scales and correspond roughly to Eysenck's 3-factor personality structure (Hampson 1988).

Table 1. Consumer Residential Preference
by Clinician Housing Recommendation[*]

	Clinician	
Consumer	*Independent*	*Group/Supported*
Independent	29%	22%
	(31)	(24)
Group/Supported	24%	25%
	(26)	(27)

Notes: [*]Scales dichotomized near median point.
$N = 108$

Analyses

Descriptive results are presented in tables of means. We test our hypotheses using analysis of variance, multivariate analysis of variance, and repeated measures analysis of variance, as appropriate. We use multiple regression analysis to evaluate the independent influence of social background, health, and social support indicators on residential preferences. The number of cases fluctuates somewhat across these analyses due to differential retention across the four waves of data collection and the particular measures included in the analyses.

FINDINGS

Consumer desire for more independent living was often discrepant with clinician recommendations (Table 1). In fact, there was no association between the scales representing clinician and consumer housing orientations. This discrepancy has important consequences for outcomes.

Housing Preferences, Recommendations, and Outcomes

Our first two hypotheses are not supported: Neither consumer housing preferences nor clinician housing recommendations that are discrepant with the type of housing obtained predict higher rates of housing loss. However, as anticipated in our third hypothesis, the combination of baseline consumer housing preferences and clinician housing recommendations was in itself a strong predictor of residential outcomes during the project: those consumers who preferred independent living but were recommended by clinicians for group housing were more likely to experience homelessness.

Table 2 presents the relevant data for all three hypothesis tests, using three measures of residential outcomes as dependent variables: percent of project days homeless, days spent in original housing before the first move, and percent of project days spent in a psychiatric hospital. For both the total sample (bottom rows) and for comparisons within the two housing types, the greatest contrast is between the two "discrepant" categories: consumers who preferred strongly to live independently but were recommended by clinicians for group living and consumers who were more willing to consider group living but were felt by clinicians to be ready to live independently.

In the total sample, consumers who had a strong preference for independent living but were recommended for group living spent 30.5 percent of their project days homeless, while those who were more favorable to group living in spite of clinician recommendation of more independence spent less than 1 percent of their days homeless. For the two other categories in which consumer and clinician preference matched, between 7 and 8 percent of project days were spent homeless.

Table 2. Residential Outcomes by Housing Type and Consumer and Clinician Housing Preference

	Both Independent	Clinician Independent Group, Consumer Group	Clinician Group, Consumer Independent	Both Group	Total	df,F	p < =
Evolving Consumer Household						df = 3,56	
% Days Hless	3.6	.00	27.6	8.7	8.1	5.6	.002
Days in Orig Hse	358	356	170	333	319	2.4	.081
% Days PsychHsp	2.9	1.2	4.6	3.7	2.9	1.2	.306
N	15	17	10	18	60		
Independent Living						df = 3,44	
% Days Hless	13.1	2.3	32.5	3.2	14.9	4.6	.007
Days in Orig Hse	294	412	273	224	297	1.6	.210
% Days PsychHsp	4.0	1.6	3.2	10.5	4.5	3.4	.026
N	16	9	14	9	48		
Total						df = 3,104	
% Days Hless	8.5	.8	30.5	6.9	11.1	10.5	.000
Days in Orig Hse	325	376	230	297	309	2.4	.071
% Days PsychHsp	3.5	1.3	3.8	6.0	3.6	2.8	.041
N	31	26	24	27	108		

Notes: MANOVA Multivariate Test: Clin X Consumer Preference, $F = 4.26$, $df = 9,300$, $p < .001$.(ECH_IL NS;ECH_IL*ClinXCons NS).
MANOVA Between-Subjects Effects for Clin X Consumer Preference (ECH_IL, ECH_IL*ClinX-Cons NS).
% Days Homeless: $F = 9.65$, $df = 3,100$ $p < .001$.
Days in Original Housing: $F = 2.99$, $df = 3,100$, $p < .05$.
% Days Psych Hospital: $F = 4.19$, $df = 3,100$, $p < .01$.

Although weaker (and not statistically significant), the same pattern occurred with days spent in original housing before a move. The pattern changed somewhat with percent of days spent in a psychiatric hospital: those consumers who agreed with the clinicians that they should live in a group home spent a higher percentage of their days hospitalized ($X = 6.0$) than consumers in the other three groups.

Consumers randomized to group housing spent a smaller percentage of their days homeless (8.1 v. 14.9) and in a psychiatric hospital (2.9 v. 4.5) than those assigned to independent apartments, although the differences were not statisti-

Table 3. Behavioral Indicators at 18 Months by Housing Type and Consumer and Clinician Housing Preference

	Both Independent Group	Clinician Independent Group, Consumer Group	Clinician Group, Consumer Independent Group	Both Group	F^2	df	$p < =$
Evolving Consumer Household							
Substance Abuse[1]	.8(15)	.9(17)	1.5(10)	1.0(18)			
Pos. Contacts	3.2(13)	2.4(16)	1.1(8)	3.9(16)			
Neg. Contacts	.5	.9	.1	.8			
Life Skills (N)	14	16	8	17			
Turbulence	1.4	1.4	1.8	1.5			
Self-Care	1.4	1.6	2.0	1.8	4.38	3,51	.01
Communication	1.3	1.1	1.3	1.3			
Responsibility	1.6	1.3	1.8	1.5			
Sociability	2.1	2.3	2.5	2.2			
Independent Living							
Substance Abuse	.8(16)	1.0(9)	1.2(14)	.8(9)			
Pos. Contacts	2.6(13)	2.2(9)	1.9(9)	4.0(7)			
Neg. Contacts	.8	1.0	.1	1.0			
Life Skills (N)	14	9	10	7			
Turbulence	1.4	1.3	1.6	1.8			
Self-Care	1.5	1.4	2.2	2.3	10.10	3,36	.001
Communication	1.2	1.3	1.4	1.7	4.76	3,36	.01
Responsibility	1.6	1.4	1.9	1.8			
Sociability	1.9	2.2	2.7	2.8	3.86	3,36	.05
Total							
Substance Abuse	.8(31)	1.0(26)	1.3(24)	.9(27)	2.84	3,104	.05
Pos. Contacts	2.9(26)	2.4(25)	1.5(17)	3.9(23)	3.74	3,87	.01
Neg. Contacts	.7	1.0	.1	.9			
Life Skills (N)	28	25	18	24			
Turbulence	1.4	1.3	1.7	1.6	3.31	3,91	.02
Self-Care	1.4	1.5	2.1	1.9	11.51	3,91	.001
Communication	1.2	1.2	1.4	1.4	2.81	3,91	.05
Responsibility	1.6	1.3	1.8	1.6	2.79	3,90	.05
Sociability	2.0	2.2	2.6	2.3	2.52	3,91	

Notes: MANOVA Multivariate Test of effects on Life Skills Dimensions: Clin X Consumer Preference, $F = 3.43$, $df = 15,252$, $p < .001$. (Effects of ECH_IL, ECH_IL*ClinXCons are NS).
MANOVA Between-Subjects Effects for Clin X Consumer Preference:
Turbulance: $F = 3.78$, $df = 3,100$ $p < .01$.
Self-Care: $F = 13.55$, $df = 3,100$, $p < .001$.
Communication: $F = 4.74$, $df = 3,100$, $p < .01$.
Responsibility: $F = 2.41$, $df = 3,100$, $p < .1$.
Social contact: $F = 2.65$, $df = 3,100$, $p < .05$.
1. Total in-project substance abuse score.
2. ANOVA statistics not presented when $p > .05$.

cally significant and did not occur for "days in original housing." The patterns of relations with the consumer-clinician housing preference variable were similar for the two housing measures, but the percent of days spent in a psychiatric hospital was much higher for those assigned to independent living who had agreed with the clinicians that they needed group living (10.5 v. 4.0, 1.6, and 3.2). This difference did not occur when these consumers were assigned to group housing.

Housing Preferences, Recommendations, and Functioning

The combination of consumer housing preferences and clinician housing recommendations predicted several problems in functioning during the project (Table 3), although the consumers who preferred independent living but were recommended by clinicians for group living were not consistently different, as we had predicted (hypothesis 3).

Consumers who sought at baseline to live independently but were recommended by the clinicians for group living were more likely to be identified as substance abusers during the project (a mean of 1.3 v. .8, 1.0, and .9). These same patterns occurred among consumers assigned to both housing types, although they were statistically significant only in the total sample. These patterns support our hypothesis 3.

Each dimension of consumer functioning in the Life Skills Inventory that was scored by case managers at the final, 18-month followup varied across the combinations of consumer preference, clinician recommendation, and housing type. The general pattern is that clinician housing recommendations predicted subsequent consumer functioning: Consumers who were recommended for group living subsequently functioned more poorly in terms of turbulence, self-care, communication, and responsibility, irrespective of their own housing preferences. We only found the interaction with consumer preference that we hypothesized (3) in one Life Skills dimension: Social functioning was poorest among those recommended for group living if they themselves sought to live independently ($F = 4.50$, $df = 1,93$, $p < .05$).

These associations tended to be stronger for consumers assigned to live independently than for those assigned to an ECH. Only the effect of clinician recommendation on self-care was statistically significant for consumers assigned to both housing types.

Consumer Preferences at Baseline

Regression analyses of consumers' housing preferences at baseline and of their desire for help with difficult things about independent living (a component of the consumer housing preference measure) highlight the discrepancy between consumer preferences and clinician housing recommendations (Table 4). Clinician recommendation of a staffed group home was not related independently to overall

Table 4. Regression of Baseline Preference for Staffed,
Group Home, Help with Difficult Things

Variable	Prefer Staffed, Group Home Beta	Don't Seek Help with Difficult Things Beta
Background		
Minority	.06	−.02
Age	.13	−.14
Female	−.28**	.27*
Education	−.11	−.07
Housing Type (IL)	−.09	.01
Health		
Schizophrenia	−.09	.09
Lifetime Substance Abuse	−.14	−.16
Psychiatric Symptoms	−.13	.08
Clinician Recommends Group	.04	.39**
Difficulty with Chores	.41***	−.34**
Social Support		
ISEL (perceived)	.09	−.10
N Positive Social Contacts	.24**	−.12
N Neg. Social Contacts	−.05	.19
R^2, Adj. R^2	.40, .31	.32, .21
N	100	87
Personality(MMPI Factors)[1]		
Psychotic	−.29*	.08
Hypochondriacal	.13	−.08
Introversion	−.06	−.17
R^2, Adj. R^2	.47, .33	.43, .25
N	79	69

Notes: [1]Betas for separate regressions including all the other predictors. MMPI interviews completed
with 98 of 118 subjects at baseline.
* $p < = .05$
** $p < = .01$
*** $p < = .001$

consumer preference for a staffed group home at baseline but it was related
inversely to consumer desire for help with things they found difficult doing in their
new residence (ß = .39). In other words, clinicians recommended a more supportive
environment for consumers who tended to reject support. Consumers who were
most opposed to living in a staffed group home also tended to score higher on the
psychoticism factor we derived from the MMPI (ß = .29).[1] By contrast, there was
an apparent consistency between consumers' lack of confidence in their ability to
manage the tasks of independent living and their preference for a staffed group
home (ß = .41) and for help with difficult things (ß = .34).

The baseline preference regression also indicates some effects of gender and
social support. Women were less likely to prefer group homes than were men and

preference for group homes increased with the number of sources of positive social support. Other indicators of social background, health and functioning, and social support were inconsequential for housing preferences. There were no independent correlates of desire to live with others (and so this regression analysis is not displayed).

Consumer Preferences at Follow-up

Residential preferences of those who remained housed changed markedly during the project's 18 months (Figure 1). The percentage of consumers preferring help with difficult things declined markedly among consumers living in both group homes and independent apartments, from a baseline level of more than 30 percent to about half that at 18 months. Preference for living with others also changed markedly during the project, but in different directions for consumers in the two housing types. Among those living in independent apartments, the

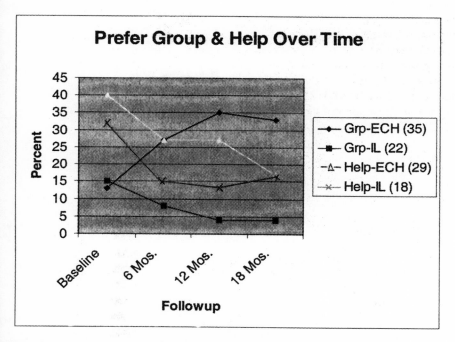

Note: Repeated measures ANOVA. Change in Help: $F = 8.68$, $df = 1,45$, $p < .05$.
Change in Group: Change X ECH-IL: $F = 2.84$, $df = 3,165$, $p < .05$.
[*]Number of Cases (those in housing for entire followup).

Figure 1. Preferences for Group Living and Staff Help Over Time

percentage preferring group living declined from 15 to 5. Among those living in group homes, however, the percentage preferring to live with others more than doubled during the first year, from just under 15 to more than 30. These changes provide strong support for hypothesis 4.[2]

After consumers had 18 months of experience in their assigned housing type, their residential preferences were not explained as well by the predictors tested at baseline (Table 5). As at baseline, consumers who felt they were less capable of doing difficult chores were more likely to prefer a staffed, group home ($\beta = .26$) and to seek help for doing things they found difficult ($\beta = .28$). However, unlike baseline, neither personality as measured by the MMPI[3] nor clinician housing recommendation was related to 18-month consumer preference. Preference for a staffed, group home at 18 months was associated independently with lower scores on our composite measure of in-project substance abuse ($\beta = -.24$).

The one social background attribute that had a significant baseline effect was still influential at 18 months: Female consumers were less interested in staffed, group living ($\beta = -.27$). Also as at baseline, the number of reported positive social contacts was associated inversely with desire for a staffed, group home ($\beta = -.24$).

Table 5. Regression of Baseline Preference for Staffed, Group Home and for Help with Things at 18 Month Followup

Variable	Prefer Staffed, Group Home (N = 88)[1] Beta	Don't Seek Help with Difficult Things (N = 81) Beta
Background		
Minority	.06	−.11
Age	.18	−.09
Female	−.27[*]	.35[**]
Education	.09	−.16
Housing Type (IL)	−.03	−.15
Health		
Schizophrenia	−.10	−.02
Subs Abuse in Project	−.24[*]	.10
Psychiatric Symptoms	−.04	−.03
Clinician Recommends Group	.17	−.05
Difficulty with Chores	.26[*]	−.28[*]
Social Support		
ISEL (perceived)	.13	−.12
N Positive Contacts	−.24[*]	.20
N Neg. Contacts	−.17	−.08
R^2, Adj. R^2	.34, .22	.26, .11

Notes: [1]Total N at 18-month followup = 90.
   [*]$p < = .05$
  [**]$p < = .01$
 [***]$p < = .001$

By the final followup, the predictors of desire for independent living had diverged for residents of the two housing types (Table 6). The residential preferences of consumers living in independent apartments were influenced more by their personal characteristics than was the case for consumers living in group homes. Among independent apartment residents, older, more educated, male minorities were more interested in living in a group home; these same characteristics had no effect among the group home residents. Those preferring group living among the independent apartment dwellers were less likely to have been diagnosed with schizophrenia ($\beta = -.48$). Substance abuse predicted a desire for independent living only among the group home residents ($\beta = -.39$), as did experience of negative social contacts ($\beta = -.30$).

The effect of clinician recommendations supports hypothesis 5 for residents of independent apartments: consumers who preferred a staffed group home tended to be those judged by clinicians as needing more support ($\beta = .37$).

Table 7 provides more information about the change from baseline to 18 months in the association between self-assessed ability to manage the tasks of

Table 6. Regression of Baseline Preference for Staffed, Group Home at 18-Month Followup by Housing Type

Variable	ECH (N = 52)	IL (N = 36)
	Beta	Beta
Background		
Minority	.09	.50**
Age	.12	.49*
Female	−.19	−.62**
Education	.00	.38*
Housing Type (IL)		
Health		
Schizophrenia	−.01	−.48*
Subs Abuse in Project	−.39*	−.23
Psychiatric Symptoms	−.17	−.15
Clinician Recommends Group	.15	.37*
Difficulty with Chores	.16	.26
Social Support		
ISEL (perceived)	.06	.18
N Positive Contacts	−.25	−.13
N Neg. Contacts	−.30*	.01
R^2, Adj. R^2	.37, .18	.58, .37
N	52	36

Notes: * $p < = .05$
       ** $p < = .01$
      *** $p < = .001$

Table 7.   Ability to Manage Difficult Things by Consumer Housing
Preference, Clinician Housing Recommendation, and
Housing Type, Baseline and 18-Month Followup

	Both Independent	Clinician Independent Consumer Group	Clinician Group, Consumer Independent	Both Group	F	df	p < =
Evolving Consumer Households							
Chores Difficult: Baseline	1.1(15)	1.2(17)	1.1(10)	1.4(18)	4.61	3,57	.01
Chores Difficult: 18 Months	1.1(13)	1.2(16)	1.4(7)	1.3(16)			
Independent Living							
Chores Difficult: Baseline	1.1(16)	1.3(9)	1.1(14)	1.2(9)	4.6	3,44	.01
Chores Difficult: 18 Months	1.1(13)	1.1(8)	1.3(9)	1.2(6)	4.6	3,32	.01
Total							
Chores Difficult: Baseline	1.1(31)	1.2(26)	1.1(24)	1.3(27)	6.6	3,104	.001
Chores Difficult: 18 Months	1.1(26)	1.2(24)	1.4(16)	1.3(22)	5.42	3,84	.05

Notes: MANOVA Test of change in ability to manage: Change by ECH_IL, $F = 4.92$, $df = 1,53$, $p < .05$.
Change by Clin X Cons Preference, $F = 4.88$, $df = 3, 53$, $p < .01$.

independent living and consumer preferences and clinician recommendations. As indicated in Table 3, at baseline consumers who sought to live independently rated themselves as more capable of handling difficult things, irrespective of clinician housing recommendations. However, at the 18-month followup, consumer self-assessment of ability to manage difficult things was higher among consumers who clinicians rated as able to live independently, irrespective of consumer housing preference (1.1 and 1.2 v. 1.4 and 1.3, $F = 5.42$, $df = 3,84$, $p < .05$). The same pattern occurred with consumers assigned to both housing types. Thus, again as anticipated in hypothesis 5, consumer housing preferences seemed at 18 months to have a more realistic basis than their preferences at baseline, before their experience in stable housing.

DISCUSSION

Consumer housing preferences and clinician housing recommendations were not associated with each other, but both helped to distinguish consumers who were likely to be more or less successful in supported or independent living. However, it was not the independent effect of preferences but rather their conjoint effect with clinician recommendations that provided the best forecast of subsequent housing outcomes and consumer functioning. The preference of consumers to live independently was as likely to indicate a rejection of help and an inability to manage the tasks of independent living as it was to indicate readiness for independent living. Clinician housing recommendations provided a basis for distinguishing consumers who fell into these two groups.

Consumers whose strong desire for independent living was discrepant with clinician recommendations were at elevated risk for subsequent housing loss and poorer functioning. However, when clinicians concurred with consumer preference for independence, residential stability was high throughout the project. In addition, consumer desire to live in a group home was a predictor of better outcomes for those consumers who were judged by clinicians to be ready for independent living.

This interaction between consumer preference and clinician recommendation had an influence on outcomes that was largely independent of housing type. The type of correspondence or discrepancy between consumer and clinician orientations was very helpful for predicting housing loss because of what it revealed about the consumers, not because it helped to identify the particular type of housing they needed. Other relationships we identified may help to explain the orientation of consumers who rejected help in spite of clinician recommendation. In particular, these consumers subsequently displayed poorer social functioning and it may have been social skills deficits that caused them to reject a staffed group home. Substance abuse was also elevated among these consumers during the project.

Clinician recommendations did help independently to predict subsequent consumer functioning, with the exception of the interaction in the case of social functioning. Thus, from the standpoint of maximizing residential stability and consumer functioning, clinician recommendations seemed "rational" while the preferences of about half the consumers were not. Considered in relation to clinician residential recommendations, consumer preference for independent or group living indicated an orientation that can be conceptualized as ranging from "treatment resistance" to help-seeking, and which was consequential for subsequent residential stability.

In the case of predicting days hospitalized for psychiatric problems, consumer preferences and clinician recommendations interacted with housing in a pattern that differed from our predictions. When consumers preferred and clinicians recommended group housing, but the consumers were actually assigned to independent living, days hospitalized were maximized. Here we speculate that consumer preference for group living may have some association with consumer willingness to agree to psychiatric hospitalization—another group living situation—or perhaps with consumer passivity. When consumers were more likely to need hospitalization due to poorer ability to manage living tasks (indicated by clinician recommendation of a group home) and the consumers were living in a situation that provided fewer alternative supports (independent living) and they were less likely to resist hospitalization (indicated by their preference for group home living), the probability of hospitalization was maximized. However, the group home environment appeared to provide enough extra support to lessen the risk identified at baseline by the clinicians. This *post hoc* interpretation should be the focus of additional research, but it suggests the importance of including hospitalization among the range of residential care options that are evaluated for consumers. The value of any particular housing alternative is determined in part by its ability to minimize the probability of hospitalization.

When the desire for someone to help with things after moving into housing was isolated from the larger housing preference measure, it was associated inversely with self-reported ability to handle chores. However, the consumers who were identified by clinicians as requiring a group home were less likely to want help after moving into housing. Thus, the more capable that consumers rated themselves, the less interested they were in help after moving into housing, but the more capable that consumers were rated by clinicians, the more interested the consumers were in help.

Changes in residential preferences among the consumers who were still in their original housing at each followup suggest that confidence in their own abilities increased in tandem with the experience of living in stable housing, whether that housing was an independent apartment or a group home. However, liking for group living changed in association with housing type, with those who experienced group living coming to like it more and those who were living alone coming to like it even less than at baseline.

The substantial change in consumer preferences with housing experience challenges the assumption of rational choice theory that preferences can be understood apart from a specific context. Consumers who remained in housing became more confident in their own abilities and increased their preference for the social situation that they experienced. Positive social contacts, which were associated with greater desire for group living at baseline, after experience were associated with desire to live independently. Social support thus seemed to take the place that theory suggests it should, as an asset to community living. These changes support a social structural interpretation of consumer preferences, but with the qualification of contingent rationality: consumers who experienced independent living seemed to base their later housing preferences on a more realistic assessment of their own abilities.

CONCLUSIONS

Since consumer preferences shape housing decisions in the open market, the extent to which mentally ill persons' preferences guide housing placements is an important indicator of the extent to which treatment of service consumers is normalized in a mental health service system. From this perspective, most mental health service systems have far to go in treating their service consumers as other citizens are treated. But from the standpoint of our findings, it appears that the goal of preference-based housing placement is appropriate for only about half of consumers like those we studied.

Homeless mentally ill individuals are not a homogeneous group and their residential preferences are not all subject to the same interpretation. For about half of our homeless consumers, the strong desire at baseline to live independently foretold less ability to maintain housing. From the standpoint of the consumers, the preference for independent living had a rational basis, as it was associated positively with perceived ability to manage the tasks of independent living. However, this confidence proved ill-founded. More self-confidence of this type predicted poorer outcomes. Our clinician raters, on the other hand, were able to make relatively sound judgments of consumers' subsequent need for support.

A central tenet of the health belief model is that interest in help for health problems will increase with the perceived severity of the disease for which help is needed. This tenet was not supported by our analysis at baseline, as indicators of health problems were not associated positively with a desire for supported living. This calls into question the applicability of rational choice assumptions. Instead, as predicted by a social structural perspective on help-seeking, both the distribution of housing preferences and their predictors changed with experience. By the end of our 18-month project, consumers who had lived independently seemed to take into account problems of their health and functioning in their preference for group or independent living. This change over time supports the hypothesis of contingent rationality: the way in which consumers understood residential situations was situated within consumer experiences (Lovell and Cohen 1998).

Substance abuse creates an exception to this conclusion. Consumers who abused substances were less likely to desire group homes after housing experience, even though they are more at risk of housing loss (Goldfinger et al. 1999). Dually diagnosed individuals may require both special recruitment approaches and unique housing options in order to reduce this risk.

The greater willingness of consumers to live with other service consumers after they had gained experience with group living is an indication that the marked discrepancy between consumer and clinician preferences identified in most surveys need not be taken as a barrier to the success of efforts to develop supportive group homes. Although we do not know whether the attitudinal changes we observed depended on the type of group housing we provided, we suspect that the basic rationale for the change was that identified by Segal and colleagues (1991):

consumer acceptance of other mentally ill persons increases with their own iden-
tification-enhancing experiences with persons who are mentally ill. The engage-
ment of consumers in the management of our group homes provided many such
experiences (Goldfinger et al. 1997).

We suspect that our central findings from the baseline analysis may apply only
to homeless mentally ill persons. Perhaps because our consumers were not housed
at baseline, their perspective on what was required to live independently was less
realistic than it became after 18 months of residential stability. Other consumers
who are already housed may not have such a sharp disjuncture between their
desires and abilities, but of course from a programmatic standpoint it is when consumers are homeless or on the verge of release from a hospital that it is most
important to define housing options and assess housing preferences. We also do
not know how experience in more traditional group homes, sometimes accused of
"infantilizing" their residents, might affect the subsequent formulation of residential preferences, but we suspect that there would not be the shift toward favoring
group living that we found in our Evolving Consumer Households.

It is also important to recognize the ambiguity inherent in tests of a rational
choice perspective, since what is deemed "rational" varies with the perspective of
the interpreter. One instance of this is the preference of women at both baseline
and 18 months to live independently. It may be that this reflected discomfort at the
prospects of (or in experience with) living in a largely male household, rather than
any disinterest in a more supportive residence. The association of psychotic
symptoms at baseline (measured with the MMPI) with a desire for independent
living may have accurately reflected the extent to which individuals suffering
from schizophrenia find group contexts to be overstimulating (Falloon and Marshall 1983), instead of indicating a disinterest in support. Substance abusers who
may have rejected staffed homes in order to maximize their drinking or drugging
opportunities were not acting "irrationally" from their own perspective. We thus
emphasize that we justify our conceptualization of rational choice strictly in terms
of the goal of maximizing residential stability.

Our methods have not included the range of measures and comparisons that
would be necessary for a convincing test of these and other implications and
complications of rational choice and social structural models of decision-making. Nonetheless, our results are consistent with the cautions of those who have
extended rational choice models and proposed alternatives to them. To para-
phrase Stoner (1985, p. 44), housing preferences emerge from a continual
reevaluation, over time, of the need for particular supports and the efficacy of
prior housing experiences. By examining housing preferences at two points in
time and among two consumer groups with different housing experiences, we
have identified some of the possibilities for such reevaluation. These same findings support Becker's (1996) emphasis on the importance of experience in
shaping preferences.

Further exploration of social structural explanations of housing preferences could reconceptualize these preferences as part of a larger strategy for achieving residential and related health and life style goals. Our results only suggest the importance of the range of alternatives, as substance abusers in some circumstances sought to live independently in part, we suspect, to facilitate their abuse. Testing Pescosolido's (1992) model also requires more sophisticated measures of social network structure and taking into account the preferences of others within the network.

Our results also highlight the inadequacy of self-report measures in the investigation of consumer functioning. Although our measure of ability to manage the tasks of daily living involved a relatively straightforward checklist, consumer responses to it suggested that their own perceptions of their abilities were at odds with their actual abilities. Since this did not seem to be true at followup, we simply point to the dependence of measurement validity on social context. We also encourage the development of more situationally specific measures of housing preferences, along the lines recommended by Goldman and colleagues (1995).

The "misfit" we have identified between consumers' housing preferences and those housing models which would best preserve their residential stability should also guide development of more appropriate housing options. Residential models must be developed which more closely meet both what consumers say they want and what clinicians believe they need. Given the strong inclination of consumers to choose independent apartments, and of clinicians to assess them as needing on-site services, we need to explore models which address both privacy needs and clinical support. Further research into programs providing both individual apartments and round-the-clock staffing should help to determine whether this combination, which *we* did not provide, maximizes the potential for maintaining residential stability. Efforts to provide active substance abuse treatment to at-risk consumers placed in housing should also be studied for their ability to reduce the rate of housing loss.

We were impressed with the predictive ability of our clinicians' ratings in terms of both residential stability and consumer functioning, but we caution against overgeneralization of this impression. Our clinician raters were experienced in housing placement and reviewed relatively substantial clinical records and interviewed consumers before making their recommendations. We presume that some such expertise and information is required for valid clinician ratings.

Clinician recommendations should be further researched with a representative sample of clinicians in order to explore the generalizability of our findings. Identification of the bases of these recommendations could help to refine training models and identify the characteristics of clinicians as well as of consumers that are associated with recommendations for more or less independent placements.

The results of this study strongly suggest that regardless of their initial housing desires, consumers tend to find whatever housing situation is offered more desirable after experience living there. This response also warrants further research, including investigation of the time frame and specific dimensions of such increased acceptability. Work which compares the impact of "on-site tours," "trial placements," and actual housing experience on residential preferences would help to clarify the programmatic implications of this potential for change (see Lovell and Cohen 1998). Options for moving between different possibilities on a residential continuum should also be evaluated for their impact on the fit between consumers' preferences and their support needs.

ACKNOWLEDGMENTS

An earlier version of this paper was presented at the 1998 Annual Meeting of the American Sociological Association, San Francisco. Research funded by NIMH McKinney Grant No. 1R18MH4808001, Stephen M. Goldfinger, MD, Principal Investigator. We are grateful for the research contributions of George Tolomiczenko, Ph.D., Win Turner, Ph.D., Barbara Dickey, Ph.D., Walter Penk, Ph.D., and Mark Abelman, M.S.W. We thank Joseph Morrissey, Ph.D., for his comments.

NOTES

1. Our confidence in these findings must be weakened due to the substantial number of cases ($N = 28$) without MMPI data.

2. Subject attrition across the followups, particularly apparent on the "help" measure, must be taken into account when assessing the likely generalizability of these findings.

3. MMPI scores were collected at 18-month followup from only 53 consumers. Neither these personality scores nor those obtained at baseline for a larger sample were associated with the categorical measure of consumer housing preference and clinician recommendation.

REFERENCES

Anthony, W. A., and A. Blanch. 1989. "Research on Community Support Services: What Have We Learned?" *Psychosocial Rehabilitation Journal* 12: 55-81.

Barrera, M. 1981. "Social Support in the Adjustment of Pregnant Adolescents: Assessment Issues. Pp. 69-96 in *Social Networks and Social Support*, edited by B. H. Gottlieb. Beverly Hills, CA: Sage.

Becker, G. S. 1996. *Accounting for Tastes*. Cambridge, MA: Harvard University Press.

Breakey, W. R. 1987. "Treating the Homeless." *Alcohol Health & Research World* 11: 42-46, 90.

Burt, R. S. 1982. *Toward a Structural Theory of Action: Network Models of Social Structure, Perception, and Action*. New York: Academic Press.

Carling, P. J. 1990. "Major Mental Illness, Housing, and Supports: The Promise of Community Integration." *American Psychologist* 45: 969-975.

Cohen, S., R. Mermelstein, T. Kamarck, and H. M. Hoberman. 1985. "Measuring the Functional Components of Social Support." In *Social Support: Theory, Research and Applications*, edited by I. G. Sarason and B. R. Sarason. Boston: M. Nihjoff.

Coleman, J. S. 1990. *The Foundations of Social Theory*. Cambridge: Belknap.

Dixon, L., N. Krauss, P. Myers, and A. Lehman. 1994. "Clinical and Treatment Correlates of Access to Section 8 Certificates for Homeless Mentally Ill Persons." *Hospital and Community Psychiatry* 45: 1196-1200.

Drake, R. E., and D. A. Adler. 1984. "Shelter Is Not Enough: Clinical Work with the Homeless Mentally Ill." Pp. 141-151 in *The Homeless Mentally Ill*, edited by H. R. Lamb. Washington, DC: American Psychiatric Association.

Eraker, S. A., J. P. Kirscht, and M. H. Becker. 1984. "Understanding and Improving Patient Compliance." *Annals of Internal Medicine* 100: 258-268.

Etzioni, A. 1988. *The Moral Dimension*. New York: Free Press.

Falloon, I. R. H., and G. N. Marshall. 1983. "Residential Care and Social Behaviour: A Study of Rehabilitation Needs." *Psychological Medicine* 13: 341-347.

Fjellman, S. M. 1976. "Natural and Unnatural Decision-making: A Critique of Decision Theory." *Ethos* 4: 73-94.

Frank, R. G., and M. S. Kamlet. 1989. "Determining Provider Choice for the Treatment of Mental Disorder: The Role of Health and Mental Health Status." *Health Services Research* 24: 83-103.

Goering, P., D. Paduchak, and J. Durbin. 1990. "Housing Homeless Women: A Consumer Preference Study." *Hospital and Community Psychiatry* 4: 790-794.

Goldfinger, S. M., and R. K. Schutt. 1996. "Comparison of Clinicians' Housing Recommendations and Preferences of Homeless Mentally Ill Persons." *Psychiatric Services* 47: 413-415.

Goldfinger, S. M., R. K. Schutt, L. J. Seidman, W. Turner, and G. Tolomiczenko. 1996b. "Self-Report and Observer Measures of Substance Abuse, in the Cross-Section and Over Time." *Journal of Nervous and Mental Disease* 184: 667-672.

Goldfinger, S. M., R. K. Schutt, G. S. Tolomiczenko, L. J. Seidman, W. E. Penk, W. M. Turner, and B. Caplan. 1999. "Predicting Homelessness After Rehousing: A Longitudinal Study of Mentally Ill Adults." *Psychiatric Services* 50: 674-679.

Goldfinger, S. M., R. K. Schutt, G. S. Tolomiczenko, W. Turner, N. Ware, W. E. Penk et al. 1997. "Housing Persons Who Are Homeless and Mentally Ill: Independent Living or Evolving Consumer Households?" Pp. 29-49 in *Mentally Ill and Homeless: Special Programs for Special Needs*, edited by W. R. Breakey and J. W. Thompson. Amsterdam: Harwood Academic Publishers.

Goldfinger, S. M., R. K. Schutt, W. Turner, G. Tolomiczenko, and M. Abelman. 1996a. "Assessing Homeless Mentally Ill Persons for Permanent Housing: Screening for Safety." *Community Mental Health Journal* 32: 275-288.

Goldman, H. H., L. Rachuba, and L. Van Tosh. 1995. "Methods of Assessing Mental Health Consumers' Preferences for Housing and Support Services." *Psychiatric Services* 46: 169-172.

Greene, R. L. 1991. *The MMPI-2/MMPI: An Interpretive Manual*. Boston: Allyn and Bacon.

Greenley, J. R., and D. Mechanic. 1976. "Social Selection in Seeking Help for Psychological Problems." *Journal of Health and Social Behavior* 17: 249-262.

Hajema, K. J., R. A. Knibbe, and M. J. Drop. 1999. "Social Resources and Alcohol-Related Losses as Predictors of Help Seeking Among Male Problem Drinkers." *Journal of Studies on Alcohol* 60: 120-129.

Hampson, S. E. 1988. *The Construction of Personality: An Introduction* (2nd ed.). London: Routledge.

Horwitz, A. 1977. "Social Networks and Pathways to Psychiatric Treatment." *Social Forces* 56 (1): 86-105.

Kadushin, C. 1969. *Why People Go to Psychiatrists*. New York: Atherton.

Keck, J. 1990. "Responding to Consumer Housing Preferences: The Toledo Experience." *Psychosocial Rehabilitation Review* 13 (April): 51-58.

Kellerman, S. L., R. S. Helpler, M. Hopkins, and G. B. Naywith. 1985. "Psychiatry and Homelessness: Problems and Programs." Pp. 179-188 in *Health Care of Homeless People*, edited by P. W. Brickner, L. K. Scharer, B. Concanan, A. Elvy, and M. Savarese. New York: Springer.

Kulka, R. A., J. Veroff, and E. Douvan. 1979. "Social Class and the Use of Professional Help for Personal Problems: 1957-1976." *Journal of Health and Social Behavior* 20 (March): 2-17.

Longshore, D., C. Grills, M. D. Anglin, and A. Kiku 1997. "Desire for Help Among African-American Drug Users." *Journal of Drug Issues* 27: 755-770.

Lovell, A. M., and S. Cohen. 1998. "The Elaboration of 'Choice' in a Program for Homeless Persons Labeled Psychiatrically Disabled." *Human Organization* 57: 8-20.

McKinlay, J. B. 1972. "Some Approaches and Problems in the Study of the Use of Services—An Overview." *Journal of Health and Social Behavior* 13: 115-152.

Minsky, S., G. G. Riesser, and M. Duffy. 1995. "The Eye of the Beholder: Housing Preferences of Inpatients and Their Treatment Teams." *Psychiatric Services* 46: 173-176.

Morse, G. A., R. J. Calsyn, G. Allen, and D. A. Kenny. 1994. "Helping Homeless Mentally Ill People: What Variables Mediate and Moderate Program Effects?" *American Journal of Community Psychology* 22: 661-683.

Owen, C., V. Rutherford, M. Jones, C. Wright, C. Tennant, and A. Smallman. 1996. "Housing Accommodation Preferences of People With Psychiatric Disabilities." *Psychiatric Services* 47: 628-632.

Pescosolido, B. A. 1992. "Beyond Rational Choice: The Social Dynamics of How People Seek Help." *American Journal of Sociology* 97 (January): 1096-1138.

Raphael, E. E. 1964. "Community Structure and Acceptance of Psychiatric Aid." *American Journal of Sociology* 69 (January): 340-358.

Rosen, A., D. Hadzi-Pavlovic, and G. Parker. 1989. "The Life Skills Profile: A Measure Assessing Function and Disability in Schizophrenia." *Schizophrenia Bulletin* 15 (2): 325-337.

Schutt, R. K., and S. M. Goldfinger. 1996. "Housing Preferences and Perceptions of Health and Functioning Among Mentally Ill Homeless Persons." *Psychiatric Services* 47: 381-386.

Schutt, R. K., S. M. Goldfinger, and W. E. Penk. 1992. "The Structure and Sources of Residential Preferences Among Seriously Mentally Ill Homeless Adults." *Sociological Practice Review* 3: 148-156.

Segal, S. P., P. L. Kotler, and J. Holschuh. 1991. "Attitudes of Sheltered Care Residents Toward Others with Mental Illness." *Hospital and Community Psychiatry* 42: 1138-1143.

Shern, D. L., N. Z. Wilson, A. S. Coen, and D. C. Patrick. 1994. "Client Outcomes II: Longitudinal Client Data from the Colorado Treatment Outcome Study." *The Milbank Quarterly* 72: 1.

Spitzer, R. L., J. B. W. Williams, and M. Gibbon. 1989. *Structured Clinical Interview for DSM-IIIR—Patient Version (SCID-P)*. New York: New York State Psychiatric Institute, Biometrics Research Department.

Stoner, B. P. 1985. "Formal Modeling of Health Care Decisions: Some Applications and Limitations." *Medical Anthropology Quarterly* 16: 41-46.

Tanzman, B. 1993. "An Overview of Surveys of Mental Health Consumers' Preferences for Housing and Support Services." *Hospital and Community Psychiatry* 44 (May): 450-455.

RACE, ROLE RESPONSIBILITY, AND RELATIONSHIP:
UNDERSTANDING THE EXPERIENCE OF CARING FOR THE SEVERELY MENTALLY ILL

Maxine Seaborn Thompson, Linda K. George, Marvin Swartz, Barbara J. Burns, and Jeffery W. Swanson

ABSTRACT

The purpose of this study was to identify contextual factors that affect the ability of caregivers to provide necessary supports to mentally ill individuals. Context was defined here as the socially patterned arrangements of peoples' everyday lives and the social and cultural meanings attached to them. Three contextual areas of caregiver burden were explored: race, role responsibility, and relationship between caregiver and the mentally ill individual. Using a stress process model as a guide for our analysis we examined the role of contextual factors, primary stressors and social supports as predictors of several dimensions of caregiver burden: objective financial burden, subjective financial burden, and household disruptions. Analyses were

Research in Community and Mental Health, Volume 11, pages 157-185.
Copyright © 2000 by JAI Press Inc.
All rights of reproduction in any form reserved.
ISBN: 0-7623-0671-8

based on interviews with 219 caregivers of persons with severe persistent mental illness who were part of randomized clinical trial of outpatient commitment (OPC) combined with community based treatment. The findings provide evidence of the importance of environmental context in structuring different aspects of caregiver burden, in particular the influence of race and relationship with the client. Parents and spouses experienced more financial burden and household disruptions than other relationships and African American caregivers reported more subjective financial burden than whites. African American caregivers also were more tolerant of client behaviors than whites. Instrumental social support and help with the client were predictors of caregiver burden. The latter was interpreted as evidence of a support mobilization effect.

INTRODUCTION

Deinstitutionalization and the subsequent reform of civil commitment laws have led to a proliferation of mentally ill individuals who revolve in-and-out of psychiatric hospitals (Hiday 1978; Lerman 1982). Many patients who have been diagnosed with severe mental illness are young adults who have few coping skills, are typically agitated and disruptive, and depend on family supports (Talbott 1984, pp. 34-36). One assumption of the community mental health movement was that family and community supports would be available to ensure that "revolving door" patients comply with treatment plans and continue prescribed medication in the community (Johnson 1990). There is growing evidence, however, that many mentally ill individuals are not provided with the resources necessary to stabilize or improve their current functioning (Johnson 1990; Talbott 1984). As such, growing attention is being paid to the importance of families as caregivers to the severely mentally ill and to factors that might assist a family in maintaining a patient in the community. The purpose of this paper is to identify factors that affect the caregiver's experience in providing supports to a mentally ill individual. In particular, we direct our attention to understanding the influence of race, caregiver's role responsibilities, and relationship to the mentally ill individual.

To date, there is evidence that the experience of caregiving to the severely mentally ill is difficult and that it varies across social groups (Clausen and Yarrow 1955; Grad and Sainsbury 1963; Hoenig and Hamilton 1966). These researchers demonstrate that the experience of caregiving differs by social class, for parents, spouses and for those who reside with the patient, but do not *explain* the processes that connect group differences to caregiver experience. Much of the research on caregiving to the mentally ill has been descriptive and of the studies that systematically examine factors that explain these difficulties, only a few rely on a theoretical framework (Maurin and Boyd 1990; Noh and Avison 1988).[1] The absence of a theoretical framework for conceptualiz-

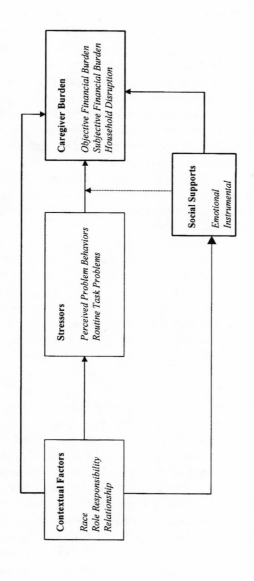

_____ Buffering effect

Source: * adapted from Aneshensel et al. 1995.

Figure 1. Social Stress Process Model of Caregiver Burden:
Duke Mental Health Study[*]

ing the processes that are related to the experience of caregiving has meant that research fails to consider how a comprehensive set of factors operate in concert to shape this experience. Those studies that do make an effort to elucidate the process of caregiving for the mentally ill rely on the *stress process* theoretical framework (Noh and Avison 1988; Pearlin, Lieberman, Menaghan, and Mullan 1981).[2] This research, however, has given considerably less attention to understanding social group variations in exposure to stress or differences in coping response. This paper addresses this issue by employing a stress process theoretical framework to examine the contextual determinants of stress among caregivers of persons with severe mental illness.

As illustrated in Figure 1 (Aneshensel, Pearlin, Mullan, Zarit, and Whitlatch 1995), the stress process model consists of four components: background or contextual factors, stressors, mediators and outcomes or caregiver burden. This model assumes that individuals in similar social statuses, social roles, and situational contexts are exposed to similar types of stressors, that the levels of resources at their disposal are similar and that the ways in which stress is expressed is also similar (Aneshensel et al. 1995, p. 35; Pearlin 1989; Pearlin, Mullan, Semple, and Skaff 1990). All too often stress researchers have neglected to examine empirically the structural origins of stressors, collecting and using such information as statistical controls (Pearlin 1989, p. 242; Pearlin et al. 1990, p. 586). In light of this omission in our understanding of the stress process, one objective of this paper is to assess whether the contextual characteristics of the caregiving experience shape and mold the caregiver's response to client's behaviors and adaptation to caregiving duties.[3]

Next, the stress model assumes that caregiving stressors (i.e., the conditions that are problematic for caregivers) drive the process that follows. Caregiving stressors stem directly from the needs of the client and the nature of the care demanded by the client (Pearlin et al. 1990). These include the problematic behaviors of the client, the daily dependencies and the surveillance and control of such behaviors. The third assumption of the model is that psychosocial resources for coping such as social supports operate as mediators of caregiving stressors. Differences in coping resources such as social supports would explain why some caregivers seem to fare better than others in similar circumstances. Another objective of this paper is to examine the contribution of social supports as mediators in the caregiving experience. Using the stress process model as a guide for our analysis (see Pearlin et al. 1990), we examine the role of contextual factors, primary stressors and social supports as predictors of caregiver burden in the public mental health sector for individuals with family members with severe and persistent mental illness (SPMI). The following section outlines in greater detail the conceptual arguments and empirical evidence for our consideration of the stress process factors.

CONCEPTUAL ARGUMENTS

Contextual Factors

How do social circumstances such as race, concurrent role obligations, and relationship between caregiver and client influence caregiver's feelings of burden? The impact of race or ethnic background on caregiver's psychological well-being rarely has been addressed, theoretically or empirically, in the existing literature on caregivers for the mentally ill (for exception, see Horwitz and Reinhard 1995b). Most research studies do not include a sufficient number of African-American or other ethnic family caregivers to allow racial or ethnic comparisons. Those that do include African-American caregivers have limited the analyses to comparison of racial differences, ignoring the linkages between race and other factors in the caregiving experience. The few studies that do include culturally diverse groups find that African-American caregivers perceive less burden associated with fulfilling their responsibilities than their White counterparts (Horwitz and Reinhard 1995b, p. 146). These findings are consistent with much of the literature on family relationships and obligations which show differences across ethnic groups regarding the expectations about independence of family members. Among some groups, individuals are expected to take care of their own basic needs, especially as adults. High value is placed on autonomy and independence. In contrast, for other groups taking care of relatives is a part of the role that one plays as a family member. We expect caregivers in African-American families to find the behavior problems and dependency of family members with SPMI less stressful than those in white families.

The context of caregiving is also shaped by the roles that caregivers occupy. Routines associated with major institutionalized roles such as parenting, marriage and working, through their repetition, create social structure. When the expectations associated with these roles are threaten, they result in stress (Pearlin 1989, p. 242). Within the course of everyday activities, caregivers are often expected to enact several roles. Difficulty in meeting the obligations and expectations of multiple roles and conflicting roles is called role overload or role conflict and is a popular explanation for the high cost of caring for the elderly (Cantor 1983; Fengler and Goodrich 1979). Multiple and competing demands on caregivers' time and resources are presumed to lead to high levels of stress. This experience is common for individuals who have commitments to work roles outside the family as well as caregiving duties within the family. Both work and caregiving require focused energy and demand a high level of commitment from role incumbents. The problems at work might have more impact when one has to attend to the needs of a mentally ill family member. Another possible source of role problems occurs as a result of competing expectations among roles within a role set. For example, one might be a caregiver for a child with mental illness and partner for a spouse within the same household. Individuals who try to meet the expectations of both wife or mother *and* caregiver might find it difficult to manage the two sets of competing responsibilities. What is most important is that the responsibility of taking care of the mentally ill is shifting from the hospital to family members (Cook and Wright 1995; Johnson 1990; Talbott 1984). However, the ways in which different role patterns shape the caregiving experience have not been investigated.

The constellation of complementary relationships that allows for differential exposure to the client is another important dimension of the caregiving context. Most studies examine the association between caregiver and client as a source or caregiver burden. Here, too, research has typically examined the impact of caregiving on the emotional well-being of parents or spouses ignoring other types of family relationships. One study reported more objective burden among spouses than among parents, but parents report more subjective burden than spouses (Hoenig and Hamilton 1966). Social and cultural norms prescribe the line of support within families. For example, spouses are obligated to take care of each other, parents take care of children, adult children are caregivers of elderly parents and so on. Role expectations and helping behaviors are not as clearly defined for other relationships such as siblings. In an exploratory study of sibling caregivers for the severely mentally ill, Horwitz (1993, p. 331) finds that brothers and sisters often step forward and provide support when parents are no longer available and, therefore, proposes the typical hierarchy of caregivers are parents and spouses, with children and siblings operating as a reserve force.

A competing explanation for differences in the experience of caregiver burden for the different types of caregiver-client relationships is the variation in their frequency of contact with the mentally ill member. Difficulties with caregiving might be more problematic for some relationships because they entail more intimate contact. Residing in the same household with the client affords opportunity for the most intimate and frequent contact. Interaction or contact with relatives can be assessed by place of residence. In the caregiving literature, studies find that living in the household with the client has more severe impact on family members: burden was higher for those caregivers who lived in the same household with their mentally ill family member. Having a family member with severe mental illness living in the home disrupted family life and placed increased strain on financial resources and day to day activities (Grad and Sainsbury 1963; Anderson and Lynch 1984; Jacob et al. 1987).

Stressors

In addition to focusing on the contextual effects on caregiver burden, the stress process model also directs our attention to an analysis of the influence of contextual factors on stressors. Both patient's illness behaviors and dependency are considered to be the primary stressors associated with the caregiving experience. Although researchers examine the impact of caregiving stressors on some measure of caregiver adaptation, they ignore the important question which is how are caregiving stressors distributed among different groups of caregivers. Past research has treated the difficulties of caring for the mentally ill as if they were an inevitable or ubiquitous characteristic of caregiving, ignoring the fact that some caregivers can manage more or less well. We argue that caregiving stressors do not occur in a vacuum but are linked to social statuses (Aneshensel 1992; Pearlin 1989, p. 245) such as race, roles, and relationships.

A summary of research findings shows that illness related behaviors and attitudes of the client (Biegel, Sales, and Schulz 1992) are related to caregiver's level of anxiety. Grad and Sainsbury's (1963) study on burden of family members of persons with mental illness in England found that anxiety was related to client symptoms such as aggression, delusions, hallucinations, confusion, and inability to care for self. Other research reports that caregivers have concerns about clients' odd speech, noisy behavior, or night wandering, and threat of suicide (Davis, Dinitz, and Pasaminick 1974; Pasaminick, Scarpitti, and Dinitz 1967; Thompson and Doll 1982). Frequency of client behavioral problems, such as sleeping problems, suspiciousness, or forgetfulness are also related to increased caregiver burden (Biegel, Milligan, Putman, and Song 1994).

Social Supports

The stress model considers social supports as mediators of primary stressors and contextual factors on caregiver strain. Social support is widely used to refer to the protective influence of interpersonal relationships for health. Social supports either intervene between stress and illness or have interactive or buffering effects that moderate the impact of stressors on burden. Several important reviews of the literature on social support provide evidence of its important role in the stress process (Cohen and Syme 1985; Cohen and Willis 1985; Gottlieb 1985; House, Landis, and Umberson 1988; Sarason and Sarason 1985). Although there is quite a proliferation of research on the nature of social support for families under stress, only a few studies have examined the association between social support and caregiver burden for families with an SPMI member. There are several difficulties in drawing conclusions from these studies. First, research studies tend to be based on nonrepresentative samples of respondents from self-help groups or family support groups (Norbeck, Chaftez, Skodol-Wilson and Weiss 1991; Potasznik and Nelson 1984; Solomon and Draine 1995). Second, most studies are descriptive or correlational (Eakes 1995; Jones, Roth, and Jones 1995). They do not identify the process by which support influences caregiver burden. For example, does social support buffer the influence of stressors or does it have an independent effect on caregiver burden? Lastly, conceptualizations of social support are usually unidimensional, referring to perceived support or emotional support (Buchanan 1995; Potasznik and Nelson 1984). Studies often ignore instrumental help with the patient in spite of the fact that family members typically identify such assistance as helpful. As Figure 1 suggests, we examine the direct, mediating and interactive effects of social supports on caregiver burden. Here, we hypothesize that social supports will ameliorate the effect of caregiving stressors on caregiver burden.

The Outcome: Caregiver Burden

Virtually all researchers agree that the experience of caring for a family member with mental illness requires caregivers to place their needs and wishes after those of patients. The term caregiver burden has been used to refer to a wide array of problems experienced by family members caring for an impaired member including physical, psychological or emotional, social and financial problems (George and Gwythen 1986; Grad and Sainsbury 1963; Hatfield 1978; Hoenig and Hamilton 1966; Thompson and Doll 1982). Both objective and subjective aspects of caregiver burden have received wide attention in the literature on mental illness. Objective burden relates to disruptions of family life caused by the illness and/or the time and effort required for the caregiver to attend to the needs of the severely mentally ill family member. It includes restriction of daily activity and time, increased expenditures of resources, and so on. Subjective burden, on the other hand, pertains to the amount of felt strain and emotional distress experienced by the caregiver (Hoenig and Hamilton 1966). Another dimension of caregiver burden is financial. Research which focuses on economic assistance, both moneys spent and in-kind services (Clark 1994), finds that the cost of caregiving is especially prohibitive for families of adult patients with severe mental illness. Thus, there is diversity in conceptualization of caregiver burden.

We hope to contribute to the theoretical development of research on caregiving for persons with a severe mental illness by applying the stress process model to an understanding of the contextual determinants of caregiving stressors and caregiver burden.

STUDY DESIGN AND METHOD

Data for this paper are drawn from an ongoing study examining the effectiveness of involuntary outpatient commitment and case management in reducing noncompliance with psychiatric treatment and preventing relapse among people with severe mental disorders. The study is a randomized clinical trial of outpatient commitment (OPC) combined with community-based treatment in a sample of more than 300 persons with severe mental disorder (Swartz et al. 1995). Since the present paper includes only the baseline data from a longitudinal study, the random assignment of subjects after their baseline interview was not an issue here. The baseline data were analyzed as one sample.

Involuntarily admitted patients were recruited from the admissions unit of a regional state psychiatric hospital and three other inpatient facilities serving the catchment area in North Central North Carolina in which the participating area mental health programs are located. Eligibility criteria for the study are (1) age 18 and older; (2) diagnosis of a psychotic disorder or major affective disorder; (3) duration of disorder for one year or more; (4) North Carolina Functional Assessment Scale score of 90 or greater (indicating significant functional impairment in activities of daily living); (5) intensive treatment within the past two years; (6) a resident of one of nine counties participating in the study; (7) a waiting a period of court-ordered outpatient commitment; (8) a clinician treatment team allows the research team to approach the patient for consent to participate. Clients with a primary diagnosis of personality disorder, psychoactive substance use disorder, or organic brain syndrome in the absence of a primary psychotic or mood disorder were excluded from the study.

Potentially eligible patients were identified from daily hospital admission records and discussion with treatment team members. While these patients were still hospitalized awaiting their outpatient commitment hearing, a research staff member met with them to describe the study and to obtain their consent for participation. Subjects were offered three benefits for participation: improved access to services via case management; possible random assignment to release from outpatient commitment orders; and monetary remuneration for each follow-up interview after hospital discharge. Of the identified patients, approximately 12 percent refused.

An extensive face-to-face interview was conducted with each respondent and with a designated family member or other collateral informant. Interviews covered a wide array of personal history information, sociodemographic characteristics, clinical history, and service use.

SAMPLE CHARACTERISTICS

Client Characteristics

Table 1 shows selected demographic characteristics of the sample. As with other studies of adults with severe mental disorders (SMD) in the public mental health system, these respondents are predominantly male younger adults, of low educational attainment, and not married or cohabiting. Perhaps more unique to this sample from North Carolina, the racial distribution is about two-thirds African American and one-third white. While a majority of clients are city residents, a substantial proportion live in rural areas and small towns.

With respect to diagnostic distribution, Table 1 shows that the sample is made up predominantly of persons with psychotic disorders; nearly 70 percent had diagnoses of schizophrenia, schizoaffective, or other psychotic disorders. An additional 27 percent had an admitting diagnosis of bipolar disorder, while only a small minority (4.5 percent) had a diagnosis of depression.[4] While current

Table 1.   Selected Characteristics of Clients and Caregivers
in the Duke Mental Health Study

Characteristics	Clients (N = 331)	Caregivers (N = 219)
Race		
African-American	66.0	62.8
White	33.0	35.5
Other	1.0	0.8
Education		
Less than High School	34.8	48.6
High School	56.6	14.4
College	8.6	37.0
Age		
18-29	17.0	9.1
30-34	51.0	25.5
45-64	29.0	43.3
65+	2.7	22.1
Gender		
Male	53.8	19.2
Female	46.2	80.8
Marital Status		
Married/co-habitating	20.0	51.4
Single	80.0	48.6
Residence		
Urban	62.0	57.7
Rural	38.0	42.3
Diagnosis (primary)		
Schizophrenia	39.0	–
Schizo affective	16.4	–
Other psychotic	14.0	–
Bipolar	27.0	–
Depression	4.0	–
Relationship to client		
Parent/grandparent	–	45.7
Adult children	–	11.5
Siblings	–	20.2
Spouse	–	8.7
Other people	–	13.9
Live with client		
Yes	–	48.0
No	–	52.0

analysis uses discharge diagnoses, approximately one-third of the sample were administered the Structured Clinical Interview for the DSM-III-R (Spitzer 1981). These interviews showed a very high level of agreement with chart review diagnoses utilizing all sources of available data.

Caregiver Characteristics

At the end of the client interview, each respondent was asked to identify a person "who knows you well, or who cares about you or helps you—anyone that you think would be a good person for us to talk with about you." A total of 289 caregivers were eligible for interviews and 266 were identified and successfully contacted. This represents a 92 percent response rate.[5] In order to estimate the influence of selection bias of informant status on the variables in the model, clients with a caregiver or informant interview were compared to those without an informant interview on several relevant characteristics. Information on the client's age, race, sex, and admission diagnosis was used for comparison. Overall the findings show no difference between clients with informant interviews and those without on these selected criteria. The analysis for this paper is based on a sample of respondents for whom there is complete data on all variables ($n = 219$).

A social profile of the caregiver sample reveals that the group is mostly female (81 percent). Of the 219 caregivers 46 percent are parents, stepparents or grandparents of the client, 12 percent are the client's adult children, 9 percent are clients' spouses, 20.2 percent are their siblings and 13 percent are other individuals. Fifty-two percent of caregivers lived with a client prior to coming to the hospital or any time during the four months prior to the client's hospitalization. The age range is 20 to 70, with an average age of 51 years old. Nearly half of the caregiver respondents are married, 17 percent are widowed, and approximately equal proportions are divorced or never married. Two-thirds of the caregivers are African American and a substantial portion had attended college and/ or received a college degree (36.8 percent).[6]

MEASURES

Dependent Variables

There are three indicators of *caregiver burden:* objective financial burden, subjective financial burden, and objective burden. *Financial burden* refers to the number of household financial contributions made on behalf of the client. Respondents were asked if they paid or provided for any of the following: transportation, pocket money for activities and recreation, clothing, food, rent, household property damaged by the client, fines or damages for others' property, legal expenses, medication, mental health treatments, medical or dental expenses, health insurance payments, time lost at work and other expenses. Each item was coded "0" for no and "1" for yes. The index scores range from 0 to 11 with a mean of 3.41 (alpha = .79). See Table 2 for the means and standard deviations of variables in the model.

Table 2.   Means and Standard Deviation of Variables
in the Analysis ($N = 219$)

Variable Name	Means	Standard Deviation
Caregiver Burden		
Objective Financial Burden	3.41	2.60
Subjective Financial Burden	4.76	2.22
Household Disruptions	6.55	4.26
Contextual Variables		
Caregiver's Race-		
(African-American)	.63	.48
Marital Status (Married = 1)	.51	.50
Working part-time	.15	.36
Working full-time	.42	.50
Co-residence	.52	.50
Parent	.46	.50
Spouse	.09	.28
Sibling	.20	.40
Adult Child	.12	.32
Stressors		
Problem Behaviors	19.31	8.87
Routine Task Problems	6.58	4.79
Social Supports		
Emotional Social Support	24.81	2.94
Instrumental Social Support	9.61	2.78
Help with Client	.68	.97
Control Variables		
Gender (Female = 1)	.79	.40
Age	51.00	14.77
Education (High School Grad = 1)	.14	.35
Education (Some College = 1)	.37	.48
Urban Residence	.58	.50
Client's Gender (Female = 1)	.42	.49
Age of Diagnosis	28.48	21.20
Affective & Depression Disorders	.32	.47
Other Psychosis	.13	.34

Subjective financial burden is the emotional cost of assuming financial responsibility for a person with mental illness. It is measured by three items which ask
the respondent "in the four months before (client) came to the hospital how bur-
dened did you feel by the financial costs of mental health treatment (including
medications, mental health office or clinic visits and related expenses)"; "by any
financial support you provided (that is, living costs for her/him?)"; and "by any
debt you got into by providing financial support?" The response codes for each
item were "1" for not at all, "2" for not much, "3" for some, "4" for a lot. The subjective financial burden scale is a summation of these items. Scores range from 3
to 12 with a mean of 4.76 (alpha = .75).

Objective burden refers to disruptions in the household or family life caused by a person with mental illness. Caregivers were asked to indicate the frequency of disruptions during the four months prior to coming to the hospital. These included disruption of household routine, limitations on social/leisure activities of members of the household, other household members feeling neglected due to the enormous amount of attention required by the ill family member, strained family relationships and lost job opportunities for adults in the household. Response categories were "0" for never, "1" for rarely, "2" for occasionally, and "3" for often. The objective burden score is formed from a summation of these items. The scores range from zero to 15 with a mean of 6.87 (alpha = .78).

Independent Variables

Contextual Variables

The social status contextual variable in the analysis is race. Caregiver's race was coded "1" for African American and "0" for white. Two indicators of role responsibilities are marital status and employment status. A dummy variable was coded for marital status (1 = married, 0 = not married). Caregiver employment is captured by two dummy variables: working full-time (1 = yes, 0 = no) and working part-time (1 = yes, 0 = no). The unemployed category includes the retired, homemakers, the disabled and those looking for work and is the comparison category. Relationship with the client is captured by three variables representing contact and interaction with the client on an ongoing or everyday basis. There is a measure for coresidence that differentiated caregivers who lived in the household with the client prior to hospitalization (coded "1") and those who did not. We also included four dummy variables representing the relationship between caregiver and client: parent, adult child, spouse and sibling (1 = yes, 0 = no). The omitted category is friends, neighbors or other nonrelatives.

Mediating Variables

Caregiving Stressors

The caregiver's behavior and duties associated with meeting the client's needs are defined as caregiving problems or stress. Caregivers were given the opportunity to report on problem behaviors and disruptions by the client in activities of daily living. First, caregivers were asked to report on a wide range of possible client *problem behaviors* that occurred four months prior to hospitalization. The list of 19 behaviors included problems sleeping (too little, too much, or at odd hours), verbal abuse, inappropriate sexual behaviors, hallucinations, temper tantrums, running away, safety carelessness, pacing aimlessly, social withdrawal, and eating too little or overeating. The response categories were coded "0" for never, "1" for rarely, "2" for occasionally, and "3" for often. The items were summed for the behavioral problems index. Scores range from 0 to 39 with a mean of 19.31.

Caregivers were also asked how often their family member had *problems with routine tasks* such as maintaining his or her personal hygiene, taking prescribed medications, preparing meals, getting dressed, doing household chores, and making use of leisure time. Each of the fourteen items was scored "0" for never, "1" for rarely, "2" for occasionally, and "3" for often. Items were summed to form an index and scores ranged from 0 to 18 with a mean of 6.58.

Social Supports

Social support is a multidimensional concept, typically including aspects of social networks, instrumental aid, and emotional or perceived support (Cohen and Syme 1985). Following this categorization scheme, we examined three social support indicators among caregivers. Two of the social support indices come from the Duke Social Support Index that was developed for and used in the Duke Epidemiologic Catchment Area Surveys (Landerman et al. 1989). An additional measure, help with the client, was developed from the caregiver's interview.

Perceived emotional social support refers to the belief that if the need arose at least one person in the individual's circle of family and friends would be helpful. The emotional social support scale is composed of ten items that tap the respondent's feelings of satisfaction with his or her relationship with family and friends. These include whether or not the respondent feels that family and friends "understand you, feel useful to family and friends, has knowledge of what's happening with family and friends, has a definite role in the family and among friends, and can count on family and friends in time of trouble." Responses were summed across items for a total score. The higher the score, the greater the perceived satisfaction with support from friends and relatives. The alpha reliability coefficient for this sample for the emotional social support scale is 70 and the mean is 24.8.

Instrumental social support refers to tangible assistance, those things that are directly involved in solving a problem. The measure of instrumental support is a sum of 13 items that asks if family or friends ever help in any of the following ways: when sick, shopping or running errands, money for bills, housework, financial or business advice, household repairs, companionship, advice about problems, transportation, and meals. The range of scores on the instrumental social support scale was 1 to 13 with a mean of 9.61.

Help with the client, another indicator of instrumental social support, refers to having other persons available to assist and/or substitute for caregivers in meeting the needs of the client. Respondents were asked if there was anyone else who handles money for the client, reminds him or her of oral medications and regularly takes the client to the clinic or doctor's office when shots were due? Positive responses were coded "1" and summed. Scores ranged from 0 to 3 with a mean of .68.

Control Variables

Both caregiver and client demographic and social characteristics were employed as controls in the analyses. Age is measured by a continuous variable indicating age in years and a dummy variable was coded for female (1 = female, 0 = male). Two dummy variables represent caregiver's education level: high school graduate (1 = yes and 0 = no) and some college (1 = yes and 0 = no). The comparison category is less than high school degree. A dummy variable was coded "1" for urban residence and "0" for rural residence. Client characteristics included sex which is coded "1" for female and "0" for male. Age at diagnosis is measured by a continuous variable indicating age in years. This information comes from the patient's interview. There are two dummy variables for diagnosis at admission: depression/bipolar (1 = yes and 0 = no) and other psychosis (1 = yes, 0 = no). Schizophrenia and schizoaffective diagnoses constitute the comparison category.

ANALYTIC STRATEGY

A series of ordinary least-squares equations is used to assess the effects of contextual factors on caregiving stressors and caregiver burden controlling for caregiver social characteristics, client social and demographic characteristics, and the client's illness characteristics. Here we postulate a block recursive framework that is parallel in its details for each caregiver burden outcome. The results are presented in two stages. First, we begin by examining each of the caregiving stressors in the model. This equation includes contextual factors and covariates for caregiver characteristics, client characteristics, and illness characteristics. In the second part of the analysis, we examine the direct and indirect effects of the contextual factors on the caregiver burden outcomes. The first equation includes the contextual factors and control variables. The second equation illustrates the effects of the contextual factors as mediated by the caregiving stressors on caregiver burden. That is, we examine whether or not the caregiving context is molded by the social reality of the client's problem behavior and dependency needs as perceived by caregivers. In the last equation, we consider whether or not social supports mediate any of the effects of caregiving stressors on caregiver burden.

RESULTS

Looking at Table 3, we see that race has a significant effect on caregiver's reports of clients' behavior problems. African-American caregivers report fewer behavior problems than whites. Role responsibilities do not influence caregiver's perceptions of clients' behavior problems. Of the relationship to client indicators,

Table 3.   Regression of Perceived Problem Behaviors and
Routine Task Problems on Contextual Variables and Control Variables
[standardized coefficients ($n = 219$)]

	Perceived Problem Behaviors	Perceived Routine Task Problems
Contextual Variables	Model 1	Model 1
Caregiver's Race-		
(African American = 1)	$-.163^{**}$	$-.207^{***}$
Role Responsibility		
Marital Status (Married = 1)	$-.002$	$.000$
Working part-time[a]	$.044$	$.057$
Working full-time[a]	$.100$	$.257^{***}$
Relationship		
Coresidence	$.200^{***}$	$-.054$
Parent[b]	$.165^{*}$	$.221^{**}$
Spouse[b]	$.074$	$.149^{*}$
Sibling[b]	$.136$	$.116$
Adult Child[b]	$.040$	$.093$
R^2	$.169$	$.155$
R^2 (Adjusted)	$.094$	$.079$

Notes: Significance level for one-tailed test
   $^{*}p < .05$
   $^{**}p < .01$
   $^{***}p < = .001$
   Control variables are caregiver's age, gender, education, urban residence, and client's gender,
   age at diagnosis and admitting diagnosis
   [a] Comparison category is the unemployed group.
   [b] Comparison category is other friends and neighbors.

several have significant effects on clients' behavior problems. Caregivers who live with their family members report significantly more behavior problems than those who do not and parents report more behavior problems than caregivers who are friends or neighbors.

Race is a significant predictor of caregiver's report of routine task problems or dependency needs with African-American caregivers reporting fewer routine task problems than Whites. As expected, caregivers with additional role obligations or role demands were concerned about routine tasks needs. Here caregivers who work full time report more routine task problems for clients than those who are not employed. Of the relationship to client variables, parent and spouse both report more problems with daily routine tasks than friends or neighbors. Overall, contextual factors explain 9 percent of the variation in problem behaviors and 8 percent of the variation in routine task problems once control variables are considered.

Table 4 presents the results of OLS models examining the contextual effects of social status, role responsibilities and exposure to the client on caregiver burden, controlling simultaneously for other covariates, through several models. Model 1 begins with the contextual factors and subsequent models sequentially add caregiving stressors and social supports. Comparisons of these models allow us to address several questions. A comparison of models 1 and 2 illustrates how the effects of the contextual variables vary with caregiver's perceptions of client's behaviors. Models 3 and higher show how social supports mediate the effects of contextual factors and stressors on caregiver burden.

First, we will examine equations that predict financial burden. Race is not a significant predictor of objective financial burden.[7] Married caregivers have lower rates of financial burden than unmarried caregivers. Caregivers who work full time report more objective financial burden than those who are unemployed. Caregivers who live with the client report more objective financial burden than caregivers who do not. And of the family relationships, objective financial burden is greatest for parents and spouse or partners than for other relations. Contextual factors and social characteristics of caregivers explain 27 percent of the residual variation in objective financial burden once covariates are considered.

In the second equation predicting objective financial burden, perceived problem behaviors and routine task problems of clients are associated with increased objective financial burden. The effect of perceived problem behaviors on financial burden is more than twice that of routine task problems. That is, each additional behavioral problem is associated with a .25 increase in financial burden compared to .13 increment in financial burden associated with each additional need activity. Overall the amount of explained variation is increased by 27 percent and this model explains 37 percent of the variation in objective financial burden. Although the absolute size of several coefficients for contextual variables are reduced, the effects of contextual factors continue to be statistically significant. Caregiving stressors account for 27 percent of the parent effect, 20 percent of the spouse and work effects and approximately 16 percent of the effect of coresidence on objective financial burden.

The mediating effects of social supports are examined in model 3. Our hypothesis that social supports mediate the effects of contextual factors and stressors on objective financial burden is not supported by these data. Only one coefficient approaches statistical significance, however. Help with client, a measure of social support that taps into assistance with needs of the client, has a negative effect on objective financial burden. The data suggest that assistance that is specifically tailored to demands of the stressful situation may be more helpful than general instrumental support.

The model of contextual characteristics does not predict subjective financial burden as well as objective financial burden, explaining only 8 percent of the variation compared to 27 percent for objective financial burden. Several of the indicators of relationship to the client attain statistical significance. Again, parents and spouse have significantly higher subjective financial burden than other caregivers. Both perceived problem behaviors and routine task problems have significant effects on subjective financial burden. Subjective financial burden is increased by

Table 4. Regression of Caregiver Burden Outcomes on Contextual Variables, Caregiving Stressors and Social Supports [standardized coefficients reported ($n = 219$)]

Dependent Variables	Objective Financial Burden			Subjective Financial Burden			Household Disruptions			
Contextual Variables	Model 1	Model 2	Model 3	Model 1	Model 2	Model 3	Model 1	Model 2	Model 3	Model 4
Caregiver's Race (African American = 1)	-.088	-.019	-.002	.052	.108	.152**	-.095	.004	.014	.034
Role Responsibility										
Marital Status (Married = 1)	-.106*	-.106*	-.097*	.008	.008	.025	.039	.039	.059	.067
Working part-time[a]	.020	.001	.002	.023	.008	.032	-.069	-.096	-.076	-.072
Working full-time[a]	.286***	.227***	.231***	.090	.037	.049	.109	.018	.037	.033
Relationship										
Coresidence	.275***	.231***	.222***	.115***	.090	.086*	.113*	.060	.050	.054
Parent[b]	.266***	.195**	.208*	.295***	.237**	.229**	.422***	.319***	.311***	.317***
Spouse[b]	.186**	.148*	.153**	.288***	.254***	.247***	.228***	.171**	.152**	.160**
Sibling[b]	.012	-.037	-.030	.119	.081	.069	.171*	.100*	.087	.101
Adult Child[b]	-.023	-.045	-.035	.071	.052	.062	.165*	.131*	.134*	.151**
Stressors										
Perceived Problem Behaviors		.250***	.272***		.162**	.179**		.325***	.324***	.398***
Routine Task Problems		.130*	.132*		.142*	.126		.224***	.207***	.216***
Social Supports										
Emotional Social Support			.010			.032			-.084	-.060
Instrumental Social Support			-.028			-.147**			-.042	-.061
Help with Client			-.089+			-.071			.013	.294*
Help with Client by Perceived Problem Behaviors										-.328*
R^2	.333	.431	.438	.155	.217	.237	.284	.486	.495	.504
R^2 (Adjusted)	.272	.374	.372	.079	.138	.147	.220	.433	.436	.443

Notes: Significance level for one-tailed test
* $p < .05$ ** $p < .01$ *** $p <= .001$ + $.10 < p < .05$
Control variable are caregiver's age, gender, education, urban residence, client's gender, age at diagnosis and admitting diagnosis
[a]Comparison category is the unemployed group.
[b]Comparison category is other friends and neighbors.

.16 with each additional behavior problem and it is increased by .14 with each additional routine task problem. When measures of caregivers' reports of clients' behavior problems and routine task problems are controlled, the effect of the parent variable is reduced by 20 percent and the spouse effect is reduced by 12 percent. Instrumental social support significantly reduces subjective financial burden, such that each helpful assistance is associated with a .15 decrease in subjective financial burden. Including the social support measures in the model, there is a marginal reduction of the contextual effects. Social supports do not explain any of the effects of the contextual variables on subjective financial burden, although the race effect ($B = .152$; $p > .05$) is enhanced with African Americans experiencing more subjective financial burden than Whites. Note the effect of routine task problems is reduced by 11 percent when social supports are considered and is no longer statistically significant.

Our next model of caregiver characteristics explains 22 percent of the variation in household disruptions—the subjective appraisal of disrupted household life. As expected, most of the relationship to client variables are significantly related to household disruptions. To obtain a sense of the relative importance of the relationship to client connections, we compare the standardized coefficients of these variables. These comparisons reveal that parent caregivers report more household disruptions, followed by spouse, siblings and adult offspring. Furthermore, relationship to client appears to be more strongly related to household disruption than coresidence. It is important to note that the sibling effect is greatly reduced (42 percent) when caregiving stressors are controlled (model 2). Thus, the greater number of household disruptions reported by siblings might be due to the way in which the problem behaviors disrupt their household affairs. Although the perceived problem behaviors account for 24 percent of the parent effect and 26 percent of the spouse effect, these variables remain statistically significant. Model 2 explains 43 percent of the variation in household disruptions. Although both problem behaviors and routine task problems significantly increase caregiver reports of household disruptions, the effect of client problem behaviors has a much stronger effect. In model 3 the social support variables are introduced and neither measure is statistically significant.

Model 4 tests whether social supports interact with client stressors in their effects on household disruptions.[8] The results show that the social support variable, help with the client , interacts with the perceived behavior problems variable. The significant, negative coefficient for the interaction between "help with client" and behavior problems means that the effect of behavior problems on household disruptions decreases as "help with client" increases. To illustrate the interaction between "help with client" and behavior problems we estimate the effect of problem behaviors on household disruptions at three levels of "help with client"—at one standard deviation below the mean, at the mean and at one standard deviation above the mean (Jaccard, Tarrisi, and Wan 1990; Marsden 1981). The corresponding computed effects are .493, .175 and −.143. None of

Table 5. Predicted Household Disruptions by Level of
Help with Client and Perceived Problem Behaviors

	Number of Perceived Problem Behaviors		
	Low	*Moderate*	*High*
Level of Help with Client			
Low	5.063	9.436	13.811
Moderate	2.026	3.578	5.131
High	−1.012	−2.276	−3.544

the other interaction effects were statistically significant. To show the pattern of social support and problem behavior effects, we calculated predicted rates of household disruptions for low social support situations (−.26), average social support (.68) and high social support (1.65) and varying numbers of perceived problem behaviors. In calculating the predicted household disruptions, we hold other variables in the model constant at their mean values. The results are shown in Table 5 where one can see that the number of predicted household disruptions are highest in situations where help with client is lowest and perceived number of problem behaviors is high. When help with client is low and number of perceived problem behaviors goes from moderate to high, the number of household disruptions is 1.4 to 2.1 above the mean. And when social support changes to a moderate level, the number of household disruptions falls below the mean. Note that all households with high social support experience levels of household disruptions that are well below the mean. In fact where number of problem behaviors and level of "help with client" are high, the number of household disruptions is at the lowest suggesting that caregivers are able to optimize their social supports. That is, how much support a caregiver receives depends on how much is needed or required to stabilize the client's inappropriate or disruptive behaviors (i.e., the level of client's symptomatology).

DISCUSSION

Our findings contribute to research on caregiver burden in several ways. First, this study joins the growing research on caregivers of severely mentally impaired persons. Few studies have focused exclusively on family members of mentally ill persons who have a history of noncompliance or grave impairment. Our sample of caregivers is composed of family and nonfamily members who may or may not reside with the client (for exception, see Tessler and Gamache 1994). Here, too we avoid selection issues associated with samples drawn from the membership of support or advocacy groups. Second, in this study we examined the context of caregiving including living arrangements and other role obligations rather than focus exclusively on the characteristics of caregivers. The findings provide evidence of the importance of environmental context in structuring different aspects of the caregiving burden. Third, our results are consistent with other studies in observing the importance of client behaviors as a source of stress for caregivers (e.g., Gubman and Tessler 1987). Fourth, social support appears to be an important protective factor associated with lower caregiver burden. We offer an interpretation of these results below.

Contextual Variables

The results of the analysis provide support for the importance of exposure to client as a source of caregiver burden, in particular interpersonal relationships between client and caregiver. The emotional consequences of caring for a severe mentally ill (SMI) individual differ for parents, spouses or siblings. Previous research has reported that parents of adult children with severe mental illness experience considerable burden. Our study confirms previous findings that parents experience more financial and objective burden than other relations (Tessler and Gamache 1994; Tausig, Fisher, and Tessler 1992; Horwitz and Reinhard 1995a). While other research argues that this parent effect is associated with their adult child's bizarre and threatening behaviors (Cook 1988; Gubman and Tessler 1987), we find the parent effect persists even when perceived problem behaviors are accounted for. Research from family studies on non-mentally ill cohorts show that parents in general expect their adult children will be financially independent and when they are not, parents experience considerable distress (Pillemer and Suitor 1991). Older parents of adult children with mental illness often express similar concerns regarding the child's ability to live independently in the community. In addition, elderly parents probably have limited financial resources and lack the physical energy that is necessary as a primary caregiver. Adult children as caregivers, on the other hand, report less financial burden than other relations, although this effect is not statistically significant. Parents with severe mental illness are a source of household burden for adult children, however. The source of the adult child's distress is not explained solely by the parent's dependency or disruptive behavior. Although this is the primary view, our findings suggest that the way in which the adult child's life is altered by their role as provider is independent of the parent's behavior. Loss of job opportunities and limitations on leisure activities are only a few ways in which a parent might interfere with the lifestyle of an adult offspring.

The emotional consequences for a spouse are equally intense as those of parents with regards to subjective burden but somewhat less than that of parents regarding financial burden and household disruptions. Reports of financial, subjective and household disruptions are higher for spouses than for other relations. The spouse-partner relationship is likely to be compromised by the mentally ill individual's behaviors. Companionship, affirmation and sharing social, financial and household obligations may go unmet. The behaviors of their partner and meeting their day-to-day needs are emotionally taxing for the caregiving spouse. Social supports do not alter these findings. It is possible that having to meet other role responsibilities might contribute to the burden experienced by spouse caregivers. It is important to note that caregivers who have the additional responsibilities associated with working full time report more financial burden. Caregivers with additional role responsibilities of working full time have less time to provide intensive personal services to family members.

The households of siblings of persons with mental illness were more problematic than those of extended relatives and other caregivers. Not surprisingly, this is solely accounted for by clients' behaviors and dependency needs. This finding suggests that because the norms to support one's brother or sister are weaker in Western society (Horwitz et al. 1992), the client behaviors might undermine their involvement in long-term caregiving.

While other research finds that living in the same household with a family member with severe mental illness is an important predictor of objective and subjective burden (Jacob et al. 1987; Solomon and Draine 1995), our findings qualify this notion. The effects of coresidence suggest that the disadvantages of living with the client are felt primarily in terms of the financial hardship that it imposes. It is probably true that the flow of monies and resources are unidirectional from caregiver to client. Although caregivers who live with the client report more problem behaviors than those who do not, this association does not explain the effect of coresidence on objective financial burden.

What social groups are most vulnerable to the demands and responsibilities of providing assistance to persons with severe mental illness? Our findings show that race structures the caregiving experience. The finding in this study of race differences in defining stressors contributes to the emerging literature on cultural differences in informal caregiving of persons with serious mental illness (see Horwitz and Reinhard 1995b). African-American caregivers report more subjective financial burden and than White caregivers. Other analyses show that the felt financial burden is net of household income suggesting a state of relative deprivation might exist. This negates the familiar income interpretation (i.e., the tradition of low incomes in African-American communities) (Farley and Allen 1987) that is often associated with race effects. Ethnographic studies of African-American families suggest that the norm or obligation to share monies with relatives is a source of stress and prohibits any one family member from accumulating resources (Stack 1974). Thus, when African-American caregivers compare their situations to that of others in their community, they might feel somewhat disadvantaged because of the financial strain of caregiving duties. Feelings of deprivation similarly might occur when African-American caregivers compare their situations to that of white caregivers. Our findings also show that African-Americans caregivers are more involved in the care of the client than Whites (Guarnaccia and Parra 1996). This, too, is consistent with the family literature which reports ethnic differences in acceptance of dependency of household members. A norm of tolerance of dependency may be more characteristic of the African-American community whereas in the white community family members maybe expected to be more independent.

Client problem behaviors and symptoms are also less bothersome for African-American caregivers compared to whites. This leads us to consider the possibility of ethnic differences in acceptance of mental illness or tolerance of deviant behaviors (Horwitz 1995b, p. 144). Previous research has shown that African Americans perceive less stigma for mental illness than Whites (Horwitz 1995b, p. 144). This interpretation is supported by our finding that African Americans perceive symptoms such as hallucinations/delusions, tantrums and acting out as less troublesome than whites and by our finding that those who report more problem behaviors have higher levels of caregiver burden. Race differences in expectations about performance or in the acceptance of problem behaviors of the mentally ill is an important factor in understanding the impact of informal caregiving of the severe mentally ill on family members.

Caregiving Stressors

It is not surprising that caregiving stressors, in particular disruptive behaviors, emerged as more important predictors of caregiver burden than contextual variables (Gubman and Tessler 1987). Clients' disruptive behaviors, disheveled appearance and dependency were perceived as problematic and a major source of caregiver burden. For example, acting out and other inappropriate behaviors associated with severe mental illness imposed substantial financial burden on caregivers and strained household routine and relationships. Caregiver burden was associated with problems that required continued care and support. Personal grooming, housekeeping and minding needs of clients led to increased emotional investment and involvement among caregivers. The context of caregiving shapes the caregiver's perceptions of client's behaviors, however.

Social Support

The social support findings observed in these data are interesting in that they show a pattern of effects that are not typically observed in other studies. Most research studies find emotional support or perceived social support is beneficial for psychological well-being. Here, emotional social support was not a significant predictor of caregiver burden. However, instrumental social support and help with the specific needs of the client were predictors of caregiver burden. Our measures of instrumental social support included a measure of help from family and friends with general needs and a measure of help from family and friends with the caregiving tasks. The latter measure is in line with researchers who have argued for measures of social support that expressed the expected stress function of the resources (Eckenrode and Gore 1981; Gore 1984, 1985; Wills 1985). Our findings contribute to this effort. Help from friends and family with general needs such as household chores running errands, when sick and companionship seem to promote caregivers emotional well-being by reducing subjective financial burden. Social network members responded to the need for help by substituting resources or caregiver efforts. Help from family and friends that included helping the client with money matters, medications, and doctor appointments reduced financial burden and buffered the impact of problem behaviors on household disruptions. The latter finding might be interpreted as a support mobilization effect. The presence of client's needs and behavior problems was met by the caregiver's mobilization of members in their network to solve or respond to caregiving problems. These problem-solving contacts might be interpreted as an extension of the personal resourcefulness of caregivers. In the presence of a threat caregivers know who they can count on for support among their network of significant others.

The stress buffering effect of help from family and friends has important practical implications. Caregivers of SMI individuals often express the need for more assistance. Interventions targeted to caregivers and clients are needed. Beyond that, however, efforts are needed to bolster the social networks of caregivers.

In the beginning of our paper we noted that the increasing role of family members in providing care for the severely mentally ill as a result of "cost shifting" policies such as deinstitutionalization and now managed care necessitate understanding the families' experience of caring for the severely mentally ill. These results demonstrate that there is great variability in levels of caregiver burden, in the three dimensions examined, but, overall, this is a highly stressed group. Thus, the severe mentally ill clients entered the hospital from (and will presumably re-enter) strained family situations. Indeed, it is even possible that the intervention, which is intended to lower rates and durations of rehospitalization, which are seen as beneficial, less coercive outcomes for clients, could actually increase caregiver burden, relative to more conventional treatment scenarios. Because of this and the high level of burden experienced by caregivers, any adequate evaluation of innovative services for the SMI should include assessment of their effects on caregivers as well as clients. As a general caveat for the study, we are unable to link differences in caregiver burden to varying long-term trajectories of caregiving patterns. That is, current levels of burden are undoubtedly in part a result of long-term caregiving demands; in contrast, our independent variables, though important, capture only the caregiving context immediately prior to hospitalization of the client. It is unlikely that some of the unexplained variance in our outcomes is a result of this.

APPENDIX

Correlation of Dependent and Selected Independent Variables[*]

	X_1	X_2	X_3	X_4	X_5	X_6	X_7	X_8	X_9	X_{10}	X_{11}	X_{12}	X_{13}	X_{14}	X_{15}	X_{16}
X_1 Race	–															
X_2 Marital Status	-.130	–														
X_3 Working, part time	-.132	.078	–													
X_4 Working, full time	-.075	.171	-.353	–												
X_5 Co-Residence	-.067	.026	-.054	.053	–											
X_6 Parent	-.059	-.043	-.008	-.207	.276	–										
X_7 Spouse	-.089	.244	.044	.119	.273	-.294	–									
X_8 Sibling	.015	.021	-.014	.168	-.305	-.458	-.156	–								
X_9 Adult Child	-.069	-.097	.016	-.057	-.129	-.325	-.110	-.172	–							
X_{10} Perceived Problem Behavior	-.188	.055	.063	.139	.225	.039	.078	.036	-.035	–						
X_{11} Routine Task Problems	-.219	.060	.029	.215	.033	.027	.061	.048	-.003	.609	–					
X_{12} Emotional Social Support	.006	.149	.047	.056	-.130	.082	-.146	-.023	-.007	-.085	-.067	–				
X_{13} instrumental Social Support	.218	.073	.130	-.063	-.044	.000	-.065	-.095	.068	-.145	-.207	.399	–			
X_{14} Help with Client	.086	.047	-.069	.056	-.044	.060	-.040	.072	.013	.221	.207	.039	-.021	–		
X_{15} Objective Financial Burden	-.116	-.002	-.094	.280	.428	.194	.204	-.156	-.167	.415	.336	-.041	-.131	.012	–	
X_{16} Subjective Financial Burden	-.034	.091	.011	.124	.225	.089	.224	-.069	-.072	.310	.270	-.090	-.173	.005	.529	–
X_{17} Household Disruption	-.173	.125	-.029	.246	.224	.045	.171	-.024	.023	.558	.476	-.173	-.203	.115	.353	.409

Notes: [*] Control Variables are omitted from the table.

The Control Variables are gender, age, education, urban residence, client's gender, age at diagnosis, admitting diagnosis. (Schzophrenia, affective depression, psychosis)

A complete correlation table is available upon request.

ACKNOWLEDGMENTS

This work was sponsored by a grant (R01 MH48103, Marvin Swartz, Principal Investigator) from the National Institute of Mental Health and by a fellowship at the UNC-CH/Duke Program on Services Research for People with Severe Mental Disorders, which is funded through a Center grant from the National Institute of Mental Health (MH51410—Joseph Morrissey, Principal Investigator). The authors wish to thank Luther B. Otto and Catherine Zimmer for their helpful comments on an earlier version of the paper.

NOTES

1.   There is a substantial body of literature on caregiving for the elderly that relies on the stress process theoretical framework (see Aneshensel, Pearlin, Mullan, Zarit, and Whitlatch 1995; Pearlin et al. 1990). Our comments are based on a review of the literature of caregiving for the severely mentally ill.

2.   The stress process model has been used quite extensively to examine issues of caregiver adaptation for other types of chronic illness, particularly those associated with aging such as Alzheimer's disease (Pearlin, Mullan, Semple, and Skaff 1990).

3.   There is considerable discussion in the stress literature on the importance of major social statuses as variables in the variation of stress. Stress research has sought to explain variation in distress as a consequence of social status variations in exposure to potential stressors and differences in coping resources. Economic status, race, gender and family roles are major social statuses (Menaghan 1990).

4.   Rates of substance abuse comorbidity were found to be quite high—over 50 percent in this sample. A low-threshold indicator of substance comorbidity was used, following the research of Clark and Drake (1994) suggesting that persons with SMD may exhibit problems associated with minimal levels of use of alcohol or illicit drugs. Accordingly, we operationally defined comorbidity as any occasional use of psychoactive substances indicated by self report or collateral report. Specifically, the criteria used were as follows: (1) respondent self-report of using alcohol or illicit drugs at least once to several times per month during the four months preceding admission, and/or (2) reports of collateral informant (usually a family member) of "excessive" use of alcohol or illicit drug use by the respondent during the four months preceding admission. Some additional clinical information on substance abuse diagnoses was available in the hospital records for these patients. However, it is well-known that hospital-based diagnostic assessments tend to under-identify substance abuse problems in acute psychiatric populations—in part because clinicians may not systematically inquire into substance use, but also because established diagnostic criteria for addiction disorders may be too stringent to identify significant drug- and alcohol-related problems in these populations. Hence, for purposes of this study, a comorbidity index based on self-and other-reported use of substances was deemed preferable to chart diagnoses.

5.   The response rate for the sample of caregivers for the Duke Mental Health Study was calculated using the formula, 1-(Refusals)/(Sampled families-Noneligible). The noneligible cases in this study included 13 cases that could not be located because of incomplete information, moves, or no information and 29 clients who did not identify a caregiver at the baseline interview. A total of 289 caregivers were identified from the client's interview and only 23 refused the interview. The refusal rate is 1-(23/(331-(13+29))) = .920.

6.   Respondents in the "Other" race category were not included in these analyses.

7.   Income which might be considered a status characteristic predicting objective financial burden is not included in these analyses because about 32 percent of the sample had missing values on income. In other analyses the inclusion of income did not have a significant effect on objective financial burden.

8. To test the interaction hypothesis we estimated a model that took the form of: $D = b_o + b_1C_1 + b_2$ stressors $+ b_3$ support $+ b_4$SupxStressor where C refers to measures of Contextual variables and the SupxStressor indicates the cross product term between the particular support variable and a stressor measure. Separate equations are estimated for each measure of caregiving burden and the interaction terms are entered individually.

REFERENCES

Aneshensel, C. 1992. "Social Stress: Theory and Research." *Annual Review of Sociology* 18: 15-38.

Aneshensel, C., L. I. Pearlin, J. T. Mullan, S. H. Zarit, and C. J. Whitlatch. 1995. "Profiles in Caregiving: The Unexpected Career." San Diego, CA: Academic Press.

Anderson, C. A., and M. M. Lynch. 1984. "A Family Impact Analysis: The Deinstitutionalization of the Mentally Ill." *Family Relations* 33: 41-43.

Biegel, D. E., E. Sales, and R. Schulz. 1992. *Factors Affecting Burden in Caregiving of Family Members with Chronic Mental Illness: A Review of the Literature.* Cleveland, OH: Mandel School of Applied Social Sciences, Case Western Reserve University.

Biegel, D. E., S. E. Milligan, P. L. Putman, and L. Song. 1994. "Predictors of Burden among Lower Socioeconomic Status Caregivers of Persons with Chronic Mental Illness." *Community Mental Health Journal* 30 (5): 473-494.

Buchanan, J. 1995. "Social Support and Schizophrenia: A Review of the Literature." *Archives of Psychiatric Nursing* 9 (2): 68-76.

Cantor, M. H. 1983. "Strain among Caregivers: A Study of Experience in the United States." *The Gerontologist* 23 (6): 597-604.

Clark, R. E. 1994. "Family Costs Associated with Severe Mental Illness and Substance Use." *Hospital and Community Psychiatry* 45 (8): 808-813.

Clark, R. E., and R. E. Drake. 1994. "Expenditures of Time and Money by Families of People with Severe Mental Illness and Substance Use Disorders." *Community Mental Health Journal* 30 (2): 145-163.

Clausen, J., and M. Yarrow. 1955. "Paths to the Mental Hospital." *Journal of Social Issues* 11: 25-32.

Cohen, S., and L. S. Syme (Eds.). 1985. *Social Support and Health.* Orlando, FL: Academic Press.

Cohen, S., and T. A. Wills. 1985. "Stress, Social Support and the Buffering Hypothesis." *Psychological Bulletin* 98: 310-357.

Cook, J. A. 1988. "Who 'Mothers' the Chronically Mentally Ill?" *Family Relations* 37: 42-49.

Cook, J. A., and E. R. Wright. 1995. "Medical Sociology and the Study of Severe Mental Illness: Reflections on Past Accomplishments and Directions for Future Research." *Journal of Health and Social Behavior* (Special Issue): 95-114.

Davis, A., S. Dinitz, and B. Pasamanick. 1974. *Schizophrenia in the New Custodial Community.* Columbus: Ohio State University Press.

Eakes, G. G. 1995. "Chronic Sorrow: The Lived Experience of Parents of Chronically Mentally Ill Individuals." *Archives of Psychiatric Nursing* 9 (2): 77-84.

Eckenrode, J., and S. Gore. 1981. "Stressful Events and Social Supports: The Significance of Context." Pp. 43-68 in *Social Networks and Social Supports.*, edited by B. H. Gottlieb. Beverly Hills, CA: Sage.

Farley, R., and W. R. Allen. 1987. *The Color Line and the Quality of Life in America.* New York: Russell Sage Foundation.

Fengler, A. P., and N. Goodrich. 1979. "Wives of Elderly Disabled Men: The Hidden Patients." *The Gerontologist* 19: 175-183.

George, L. K., and L. P. Gwyther. 1986. "Caregiver Well-Being: A Multi-Dimensional Examination of Family Caregiver of Demented Adults." *The Gerontologist* 26 (3): 253-269.

Gore, S. 1984. "Stress-Buffering Functions of Social Supports: An Appraisal and Classification of Research Models." Pp. 202-222 in *Stressful Life Events and Their Context*, edited by B. S. Dohrenwend and B. P. Dohrenwend. New Brunswick, NJ: Rutgers University Press.

———. 1985. "Social Support and Styles of Copying with Stress." Pp. 263-277 in *Social Support and Health*, edited by S. Cohen and S. L Syme. New York: Academic Press.

Gottlieb, B. 1985. "Assessing and Strengthening the Impact of Social Support on Mental Health." *Social Work* 304 (4): 293-300.

Grad, J. P., and P. M. D. Sainsbury. 1963. "Mental Illness and the Family." *Lancet* 1: 544-547.

Guarnaccia, P., and P. Parra. 1996. "Ethnicity, Social Status and Families Experiences of Caring for a Mentally Ill Family Member." *Community Mental Health Journal* 3: 243-260.

Gubman, G., and R. Tessler. 1987. "The Impact of Mental Illness on Families." *Journal of Family Issues* 8: 226-245.

Hatfield, A. 1978. "Psychological Costs of Schizophrenia to the Family." *Social Work* 23: 355-359.

Hiday, V.A. 1978. "Mental Commitment Legislation: Impact on Length of Stay."*Journal of Mental Health Administration* 6: 4-13.

Hoenig, J., and M. Hamilton. 1966. "The Schizophrenic Patient in the Community and His Effect on the Household." *International Journal of Social Psychiatry* 12: 105-176.

Horwitz, A. V. 1993. "Siblings as Caregivers of the Seriously Mentally Ill." *The Milbank Quarterly* 7 (12): 323-339.

Horwitz, A. V., R. C. Tessler, G. A. Fisher, and G. M. Gamache. 1992. "The Role of Adult Siblings in Providing Social Support to the Severely Mentally Ill." *Journal of Marriage and the Family* 44: 155-167.

Horwitz, A. V., and S. Reinhard. 1995a. "Family and Social Network Support for the Seriously Mentally Ill: Patient Perspectives." Pp. 205-32 in *Research in Community and Mental Health*, vol. 7, edited by J. Greenley. Greenwich, CT: JAI Press.

———. 1995b. "Ethnic Differences in Caregiving Duties and Burdens among Parents and Siblings of Persons with Severe Mental Illnesses." *Journal of Health and Social Behavior* 36: 138-150.

House, J. S., K. R. Landis, and D. Umberson. 1988. "Social Relationships and Health." *Annual Review of Sociology* 14: 293-318.

Jaccard, J., R. Turrisi, and C. K. Wan. 1990. *Interaction Effects in Multiple Regression*. Beverly Hills, CA: Sage.

Jacob, M., E. Frank, D. J. Kuper, and L. L. Carpenter. 1987. "Recurrent Depression: An Assessment of Family Burden and Family Attitudes." *Journal of Clinical Psychiatry* 48: 395-400.

Johnson, A. B. 1990. *Out of Bedlam: The Truth About Deinstitutionalization*. New York: Basic Books.

Jones, S., D. Roth, and P. K. Jones. 1995. "Effect of Demographic and Behavioral Variables on Burden of Caregivers of Chronic Mentally Ill Persons." *Psychiatric Services* 46 (2): 141-145.

Landerman, R., L. K. George, R. T. Campbell, and D. G. Blazer. 1989. "Alternative Models of the Stress-Buffering Hypothesis." *American Journal of Community Psychology* 17: 625-642.

Lerman, P. 1982. *Deinstitutionalization and the Welfare State*. New Brunswick, NJ: Rutgers University Press.

Marsden, P. V. 1981. "Conditional Effects in Regression Models." Pp. 47-116 in *Linear Models in Social Research,* edited by P. V. Marsden. Beverly Hills, CA: Sage.

Maurin, J., and C. B. Boyd. 1990. "Burden of Mental Illness on the Family: A Critical Review."*Archives of Psychiatric Nursing* 4 (2): 99-107.

Menaghan, E. G. 1990. "Social Stress and Individual Distress." Research in *Community and Mental Health* 6: 107-141.

Noh, S., and W. R. Avison. 1988. "Spouses of Discharged Psychiatric Patients: Factors Associated with their Experience of Burden." *Journal of Marriage and the Family* 50: 377-389.

Norbeck, J. S., L. Chaftez, H. Skodol-Wilson, and S. Weiss. 1991. "Social Support Needs of Family Caregivers of Psychiatric Patients from Three Age Groups." *Nursing Research* 40 (4): 208-213.

Pasamanick, B., F. Scarpetti, and S. Dinitz. 1967. *Schizophrenics in the Community.* New York: Appleton-Century-Crofts.

Pearlin, L. I. 1989. "The Sociological Study of Stress." *Journal of Health and Social Behavior* 30: 241-256.

Pearlin, L. I., M. A. Lieberman, E. G. Menaghan, and J. T. Mullan. 1981. "The Stress Process." *Journal of Health and Social Behavior* 22: 337-356.

Pearlin, L. I., J. T. Mullan, S. J. Semple, and M. M. Skaff. 1990. "Caregiving and the Stress Process: An Overview of Concepts and Their Measures." *The Gerontologist* 30 (5): 583-594.

Pillemer, K., and J. J. Suitor. 1991. "Will I Ever Escape My Child's Problems? Effects of Adult Children's Problems on Elderly Parents." *Journal of Marriage and the Family* 53: 585-594.

Potasznik, H., and G. Nelson, 1984 "Stress and Social Support: The Burden Experienced by the Family of a Mentally Ill Person." *American Journal of Community Psychology* 12 (5): 589-607.

Sarason, I. G., and B. R. Sarason (Eds.). 1985. *Social Support: Theory, Research, and Applications.* Dordrecht, Netherlands: Martinus Nijhoff.

Solomon, P., and J. Draine. 1995. "Subjective Burden Among Family Members of Mentally Ill Adults:Relation to Stress, Coping and Adaptation." *American Journal of Orthopsychiatry* 65: 419-427.

Spitzer, R. L., S. Andrew, M. Gibbon, and J. B. W. Williams. 1981. *DSM-III Casebook.* Washington, DC: American Medical Association.

Spitzer, R. L., J. Endicott, and E. Robbins. 1978. "Research Diagnostic Criteria: Rationale and Reliability." *Archives of General Psychiatry* 35: 773-782.

Stack, C. 1974. *All Our Kin: Strategies for Survival in a Black Community.* New York: Harper Row Publishers.

Swartz, M., B. Burns, V. A. Hiday, L. George, J. Swanson, and N. R. Wagner. 1995. "New Directions in Research on Involuntary Out-Patient Commitment." *Hospital and Community Psychiatry* 46: 381-385.

Talbott, J. A. 1984. *The Chronic Mental Patient: Five Years Later.* New York: Grune and Stratton, Inc.

Tessler, R., and G. Gamache. 1994. "Continuity of Care, Residence, and Family Burden in Ohio." *The Milbank Quarterly* 72 (1): 149-169.

Tausig, M., G. A. Fisher, and R. C. Tessler. 1992. "Informal Systems of Care for the Chronically Mentally Ill." *Community Mental Health Journal* 28 (5): 413-425.

Thompson, E., and W. Doll. 1982. "The Burden of Families Coping with the Mentally Ill: An Invisible Crisis." *Family Relations* 31: 379-388.

Willis, T. A. 1985. "Supportive Functions of Interpersonal Relationships." Pp. 61-82 in *Social Support and Health*, edited by S. Cohen and S. L. Syme. New York: Academic Press Inc.

PART III

METHODS

STRATEGIES FOR EVALUATING PUBLIC SECTOR MANAGED CARE USING ADMINISTRATIVE DATA

Neil M. Thakur, Joseph P. Morrissey,
William H. Sledge, and Mark Schlesinger

ABSTRACT

Public sector managed behavioral care is changing faster than researchers can evaluate its effects. This chapter presents a format for managed behavioral health evaluations which is less expensive and easier to implement than methods which require prospective data collection. A framework is presented for evaluating managed care interventions based on two criteria: (1) access to care, defined as first contact with a behavioral health clinician; and (2) changes in utilization patterns. These issues can be addressed using three pre-existing electronic administrative data sources: Medicaid paid claims and enrollment rosters; state substance abuse datasets; and state or county mental health treatment datasets. Various linkage strategies are presented, as well as statistical methods that handle multiple repeated time elements, such as policy periods, changing insurance status, and episodes of care. Potential controls and comparisons are also discussed. Finally, the human subjects implications of these designs will be addressed.

Research in Community and Mental Health, Volume 11, pages 189-203.
Copyright © 2000 by JAI Press Inc.
All rights of reproduction in any form reserved.
ISBN: 0-7623-0671-8

INTRODUCTION

Virtually all states have implemented some type of Medicaid reform, and these changes have often occurred faster than they can be evaluated. Mechanic (1998) suggests that state mental health authorities adapt to this new environment by becoming "the watchguards for the privatized public safety net." Effective performance in this role will require rigorous program monitoring and evaluation. Evaluating behavioral managed care by collecting prospective data on client outcomes, perhaps the most rigorous non-experimental design, is intrusive to subjects, expensive, and sometimes not even possible due to the rapidity of managed care implementation. Existing electronic records can provide longitudinal data efficiently, but designs based on these data are hindered by methodological, technological, political, and legal complications. As more and more researchers are drawn in to managed care evaluations, there is a need for a primer on the problems and prospects of using administrative data in these evaluations. This chapter responds to this need by discussing ways to use electronic databases to conduct public sector behavioral health managed care evaluations.

Using administrative data in public sector evaluations generates a typical set of problems and advantages. Several types of information bias commonly occur: errors in linking subject records across systems; high and/or differential rates of missing data; and failure to collect important data elements such as severity of illness or incarceration. Methodological problems are also common, such as: statistical challenges in dealing with long follow-ups and uncontrolled policy interventions; the lack of experimental groups; and the difficulty in generalizing findings from essentially unique managed care programs. Finally, administrative data introduces special legal and political challenges, including accessing data from multiple service agencies with varying confidentially standards.

Despite these problems, administrative data have number of distinct advantages. They are typically more comprehensive than any other alternative. Studies using chart review and client/provider interviews must often rely on smaller samples because they are far slower and more expensive than electronic data sources. Further, retrospective data collection, a necessity in a chaotic policy environment that hinders long term research planning, becomes impractical as subjects relocate or paper service records are archived. Most important, without administrative data, it is virtually impossible to obtain population based denominators for access and utilization rates.

This chapter will suggest methods for using administrative electronic records to evaluate adult services in two major areas—access to care and substitution of care. These methods will be presented in three steps. First, a schematic for key managed care evaluation questions will be developed. Second, some of the existing electronic databases maintained by states and localities as part of their normal functions will be listed. Finally, access and substitution will be addressed in light of the practical problems discussed above.

A SCHEMATIC FOR EVALUATING
PUBLIC SECTOR MANAGED CARE

Changes in access and utilization of services that may result from managed care are illustrated in Figure 1. This schematic is not intended to be an exhaustive list of all possible managed care outcomes. Rather, it highlights the kinds of outcomes a managed care program might be expected to achieve in order to be renewed or be considered for replication in other sites.

This schematic describes two steps for evaluation. First, as indicated by the "Receipt of any care by enrollees decision point," a managed care intervention should maintain or increase enrollees' access to services to be judged successful. Though there are many dimensions of access (Gold 1998), probably the most important is the opportunity to become engaged in treatment. In this sense, access can be defined as any contact with a behavioral health professional. We propose that reducing this dimension of access is an unsatisfactory outcome for all behavioral health managed care, and if it occurs, is sufficient evidence to warrant a program correction.

Second, once access levels have been empirically determined, the quality of care must also be evaluated. Figure 1 represents the nature and amount of care substitution that occurs through the "utilization decision point." The ideal outcome is listed in far right corner, labeled "appropriate substitution." This is a synopsis of clinical and fiscal goals that are at the heart of most managed care initiatives. Success requires improvement in clinical outcomes, same or lower direct costs per client, and a reduction in cost shifting.

But assigning managed care programs to a box on a schematic is only one step in the evaluative process. To be truly useful and generalizable, a good evaluation should attribute program outcomes to specific program mechanisms—a process researchers are just beginning. The rationale underlying this schematic is explained below as an introduction to suggesting ways in which administrative data can be useful in this context.

ACCESS TO CARE: RECEIPT OF
ANY CARE DECISION POINT

Managed care's financial incentive of pre-payment is assumed to affect access in two contrasting ways (Miller 1998). First, as Mechanic and colleagues (1995) suggest, providers have powerful incentives to ignore potentially high utilizing clients, especially if utilization is higher than expected, thereby leaving the provider with a treated pool of less expensive (low-intensity) clients. In contrast, based on the substitution argument described below, others claim that the likelihood for receiving care should increase under capitation, especially for under-served groups such as people with serious mental illness.

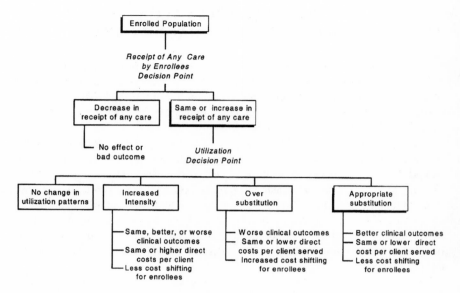

Note: This model presents some of the potential public sector behavioral health managed care outputs (listed below the boxes) that may result from a given condition (in boxes). Decision points suggest criteria to empirically determine the success of public sector managed care programs. Appropriate substitution (in the lower right corner) in the presence of the same or increased access is the only condition that will both lower cost shifting and cost per client served.

Figure 1.   A Schematic for Managed Care Evaluation

The evidence linking capitation to changes in access to behavioral health care is limited. In an evaluation of capitated services for people with serious mental illness in Arizona, Leff and colleagues (1994) found a slight increase in probability of use and appropriateness of treatment for the capitated program compared to the fee for service baseline program. In the Massachusetts capitated Medicaid carve out, Callahan and colleagues (1994) and Dickey and colleagues (1996) found no evidence of any changes in the probability of use.

In light of these conflicting views and evidence, it seems reasonable to anticipate that enrollee access to any type of care may increase or decrease in response to managed care. These outcomes are reflected under the *Receipt of Any Care Decision Point* in Figure 1. Though this decision is independent of the quality of services received, a reduction in access are considered a negative outcome so its effect on subsequent utilization is not shown on the schematic.

SUBSTITUTION OF CARE:
UTILIZATION DECISION POINT

Managed care not only affects the likelihood of a person getting care, but also the kinds of care that a person will receive. Utilization management reallocates resources by substituting high intensity services with low intensity services, and this substitution may affect quality of care. The financial arrangements discussed above may play an indirect role as well. The *Utilization Decision Point* in Figure 1 graphically represents four of the possible effects of managed care on behavioral health service utilization. Appropriate substitution, the only ideal outcome for most managed care programs, is shown on the far right.

As indicated at the bottom left of the schematic, a managed care program may have no effect on utilization patterns. Moving right, if a population were previously grossly under-served, a managed care intervention may result in increased intensity of service. Such a situation may occur when expanding eligibility or actively engaging populations into treatment that previously had little or no treatment. For example, when providing managed care services to people who are homeless, one would hope to see higher rates of service, higher direct costs per client, and reduced cost shifting from behavioral health care to prisons, primary health care, or families. Errors in formation of a managed care plan, such as inappropriate financing incentives or rigid network management, could theoretically increase use of high intensity services like inpatient care for clients. Under current notions of community based treatment (Breakey 1996; Lehman and Steinwachs 1998), unwarranted increases in inpatient services may be considered poor quality care because it could decrease the ability of a client to function in the community and would reduce the likelihood of appropriate care for other members of the population.

Over substitution, or under treating clients by relying too heavily on low intensity services, is the third potential output of the Utilization Decision Point (Figure 1). Over substitution may result in worse clinical outcomes, increased cost shifting, and potentially lower direct costs per client.

Over substitution is a real danger in many managed care plans. Though the Massachusetts Medicaid carveout substituted outpatient care for inpatient psychiatric care and inpatient detoxification (Thompson, Burns, Goldman, and Smith 1992), Mechanic and colleagues (1995) have argued that this substitution was too extreme and may have resulted in inappropriate care. Over substitution may also have occurred in the initial Arizona behavioral long term care demonstration due to fiscal pressure, as enrollment was expanded without a proportional increase in funding (Santiago and Barren 1989), and in Tennessee's failed behavioral health carveout, were Chang and colleagues (1998) linked service reductions with budget cuts, profit taking, and payment delays. In the Rand Medical Outcomes study, too few services provided by HMOs may have led clients with major depression to change insurers (Rogers, Wells, Meredith, Sturm, and Burnam 1993; Sturm et al. 1995). In Massachusetts, Stroup and Dowart (1995) found no evidence of cost shifting to its state mental health authority in an early evaluation of the Massachusetts Medicaid program. However, Dickey and associates (1996) found in a later study of the same program that savings among schizophrenic patients on inpatient care were offset by increased bed days in facilities operated by the state mental health authority.

The fourth condition, Appropriate Substitution (in the lower right corner of Figure 1), describes positive effects on the population eligible for care. It is the only condition that reduces cost shifting while maintaining or improving clinical outcomes and direct costs per client, and therefore, is the only condition that can clearly be called an improvement. This is an objectively modest standard, but there are few documented examples of it having been met in current practice. It may have been met initially in New York state's capitation demonstration, though costs did rise as the program ended. In this case, capitated clients had fewer inpatient treatment days, but had no detectable differences from fee-for-service baseline clients in functioning or levels of symptoms (Cole, Reed, Babigian, Brown, and Fray 1994; Reed, Hennessy, Mitchell, and Babigian 1994). In the Minnesota experiment, there was no significant difference between fee-for-service and capitation clients with serious mental illness in terms of general health and psychiatric symptoms (Lurie, Moscovice, Finch, Christianson, and Popkin 1992).

USING ELECTRONIC DATA SOURCES TO EVALUATE THE IMPACT OF PUBLIC SECTOR MANAGED CARE

Given the inconsistency of managed care effects reported in the current literature, there is a pressing need for evaluations to further our understanding of the impacts of managed care on access and utilization. Electronic data sources to facilitate such evaluations are described below.

Data Sources for Access Measures

As Figure 1 suggests, access is a rate, and must be judged with reference to both a numerator (persons served) and a denominator population (persons eligible for care). Insurance programs, such as Medicare or Medicaid, maintain eligibility rosters indicating when individuals have active coverage. Medicare enrollment files are available from a national database maintained by the Health Care Finance Administration (HCFA), whereas there is no central location that stores Medicaid eligibility files. Rather, each state maintains its own Medicaid data system, and passes specific elements up to the national level. Often, information on dual enrollment (concurrent Medicare and Medicare enrollment) is also recorded by state Medicaid authorities. Although difficult to disaggregate in national or multi-state studies, these enrollment episodes serve as the basis for a true, dynamic denominator population, with people entering and exiting over time.

Data Sources for Utilization Measures

The numerator for the access rate, the number of people receiving services, comes from utilization databases. Data for publicly funded behavioral health services are stored in three places—Medicaid paid claims files; publicly operated state and local mental health information systems; and publicly maintained substance abuse treatment provider utilization databases.

A subset of both mental health and substance abuse data is found on Medicaid paid claims files. These files are probably least research friendly of all of these data sources, due to their complexity, processing delays, file size, number of unrelated records, and confidentiality restrictions. More importantly, a Medicaid database cannot be used as a baseline data source for new services or populations newly eligible under the managed care intervention, since it would not have collected data on these services or people prior to the intervention. Further, not all services offered by the public sector will be covered by the new Medicaid managed care program, and Medicaid managed care firms may be able to shift costs and treatment responsibility for certain clients onto other human service systems.

Another important utilization data source are information systems operated by state or county mental health authorities. These systems typically assign a unique identifier that follows a person throughout their services in the public sector mental health system. However, utilization data may be stored in a separate database. Washington state psychiatric hospital admission and discharge data, for example, are recorded by the state on one database, but counties maintain separate databases for outpatient services provided by public clinics. In contrast, many clinic services and state hospital services in Connecticut are managed on the same information system (though it is being replaced). Under both conditions, information systems will usually either have a shared unique identifier or a pre-established method for linking client records together. As a result, the number of databases and their agency of origin are not important design issues, though they may have political/data access implications for the evaluator.

Substance abuse services provided under contract to most state substance abuse authorities are reported to Substance Abuse and Mental Health Agency's (SAMHSA) Treatment Episode Data set. They usually only include licensed substance abuse providers, and therefore do not include detoxification and emergency room services provided by general hospitals. In most cases, the latter must be extracted from the local Medicaid paid claims files. Like the Medicaid paid claims files, substance abuse datasets usually have a several month processing delay. Unlike the other databases discussed, however, these data sets do not always maintain a master client record. Rather, each episode of care is stored as a record with a unique client identifier attached, necessitating an internal linkage process to create a person-level file before it can be linked with external data sources.

PRACTICALITIES

These administrative information systems contain the necessary data to conduct client-level longitudinal evaluations of managed care interventions. Below is some advice about translating these multiple data sources into analytic files for program evaluation and research.

Dealing with Information Bias

Three types of information bias are common in administrative data studies—linkage errors, uncollected information (sometimes referred to as "omitted variables"), and high or differential rates of missing data. These problems are endemic to electronic records of service encounters, but they are not insurmountable.

Linkage Techniques

Records from multiple data systems need to be linked before they can be analyzed. There are two sets of standard techniques for doing this—deterministic and probabilistic matching (Adams et al. 1997; Alsop and Langley 1998; Bell, Keesey, and Richards 1994; Roos and Wajda 1991; Roos, Walld, Wajda, Bond, and Hartford 1996; Wajda, Roos, Laefsky, and Singleton 1991).

Deterministic matching links records that exactly match on reliable criteria or combinations of criteria (e.g. name, Social Security Number). Due to recording errors, and vagaries of names across episodes (e.g., name changes, duplicate names), deterministic matching can result in an undercount of matched records.

Probabilistic matching is a more advanced technique whereby records are linked even if they are not identical on all identifiers. Formal probabilistic procedures involve a two step process of assigning weights to potential linkage fields based on their assumed accuracy and then linking the records if the sum of the weights crosses a certain threshold.

Though some errors are likely under both deterministic and probabilistic matching, they are controllable through a variety of techniques (Brenner 1994; Brenner and Gefeller 1993). Probabilistic matching can also estimate the error rate of record linkage.

Uncollected Information

Two sets of data are typically omitted from public sector behavioral health databases. These are measures of subjects exiting a cohort, such as death or imprisonment, and omitted outcome measures, such as client satisfaction or clinical status. Suggestions for dealing with these types of omitted data are listed below.

It is very difficult to track client deaths and incarcerations in behavioral health databases. It is even harder to track people moving out of a study area. At best, one can try to fill these gaps by using other information sources such as the Social Security Death Index (Smith 1998), or obtaining anonymous population estimates of incarceration as done for studies in Vermont (Pandiani, Banks, and Schacht 1998b), or actual lists of incarcerated subjects, as done in a Texas study (Jackson, Sullivan, Kogel, Morton, Fujisaki 1998).

Another common problem is the lack of direct measures of client satisfaction. Because of their lack of availability in most administrative data systems, satisfaction was not included as an outcome category in Figure 1. However, one can measure patient *dis*satisfaction indirectly through disenrollment rates (McFarland, Johnson, and Hornbrook 1996; Schlesinger 1989; Sturm et al. 1995; Sturm et al. 1994).

The most serious omitted data problem is absence of clinical status and level of functioning indicators. Though administrative records offer few measures of symptom reduction or cure, this lack of information is as much a limitation of current clinical standards as it is of electronic data systems themselves. Standard process indicators, however, such as continuity of care, time to rehospitalization, or frequency of emergency utilization, are typically available through administrative records.

Dirty Data

There are two major sources of data integrity problems, or dirty data, that are peculiar to electronic data. They are bias caused by transactional data collection, and poor or missing diagnostic information.

Missing data in electronic records are generally the same as missing data from other sources. However, missing information about a person (rather than the care episode) is especially vulnerable to bias on transaction based data files. Each new transaction (e.g., clinical encounter) is an additional data collection opportunity for client level information (e.g., race, sex, housing status). This information is much more likely to be missing for low utilization subjects, because subjects with few clinical encounters have less opportunities to have accurate information entered into the data base than high utilization subjects (Thakur, Hoff, Druss, and Catalanotto 1998). Imputing missing client–level variables from high utilization subjects to low utilization subjects can be inappropriate if the imputed variables are associated with service use.

Another problem is the difficulty in obtaining diagnostic details from electronic administrative data. Each episode of care typically has multiple diagnostic codes, and it is up to clinicians and coders to assign their order of importance in the electronic record. Often, chart reviews, which allow for more subtle interpretations of symptoms and diagnostic combinations, can be used to supplement the electronic data. Chart review also provides a better check on the accuracy of the information included in an analysis than electronic systems which are vulnerable to mis-entry, and samples of electronic records can be verified with chart review in some circumstances. Though there are no easy solutions for this verification problem, electronic records often serve as a financial gold standard—they are the basis for most medical bills and payment.

Dealing with Methodological Problems

In addition to information bias, methodological aspects of electronic data based evaluations bear further discussion. The size and complexity of these databases raise statistical problems, but they also provide the opportunity to assemble comparison groups to explore in considerable depth the access and utilization questions raised in the schematic.

Statistical Challenges

Amalgamation of the datasets described above will result in a combined file with three repeated time elements. The first are *periods* that account for different phases of implementation of key managed care elements and/or other salient policies. A second repeated element is multiple exclusive *treatment episodes*—for example, institutionalization that precludes other concurrent forms of care. The final element is multiple *episodes of insurance enrollment*.

Longitudinal analytic techniques, such as Cox or Poisson regression with generalized estimating equations (Liang and Zeger 1986; Lipsitz and Kim 1994; Zeger, Liang, and Albert 1988), can be used to describe the changes in access and utilization over time, while accounting for these multiple repeated time elements. In cases were a specific event date is not needed (e.g., costs incurred per client over a time period) linear regression with multiple repeated elements can be modeled by manipulating covariance matrices (Galecki 1994; Jennrich and Schuster 1986). Both of these methods are computationally intensive and technically difficult, but necessary to avoid underestimation of error. Alternative methods would include simplifications such as ignoring follow-up information while insurance coverage is inactive, or running separate models for each insurance status (active insurance coverage and inactive insurance coverage).

Assembling Control Groups

Despite this increased statistical complexity, the advantage of such a comprehensive analytic database is that it allows for the creation of controls and contrasts in observational designs. Period effects can be minimized and attention can be focused on the intervention of interest by contrasting outcomes among different insurance/policy groups. Two types of insurance contrasts are typically available: state and county funded behavioral health services versus Medicaid services; and different managed care strategies within the same Medicaid program (as in Iowa with its behavioral health carveout and physician case management options)(Hoge, Jacobs, and Thakur 1997). However, insurance group contrasts alone are not adequate, because assignment to insurance groups is usually not random, and insurance enrollment changes over time.

Within-subject comparisons, especially repeated measures of utilization, allow subjects to serve as their own control for severity of chronic illnesses. The weakness of this technique is that symptoms can naturally vary simply due to passage of time (Draine and Solomon 1999). This potential bias can be minimized through joint subject and insurance contrasts, which compares one insurance group with another concurrently, and also compares individual subjects with themselves longitudinally as their insurance changes over time. These joint contrasts, often statistically modeled through interactions, are the least biased method to explore specific policy effects and minimize subject and period effects.

Sampling and Generalization

Unlike other types of evaluation designs, evaluations based on administrative electronic data allow for an entire target population to be sampled and included in a statistical analysis. Standard statistical tests where inferences are made about the sample and generalized to the target population become meaningless under these circumstances. However, if the system under study is viewed as a representative case of a category of managed care programs also found in other sites (or even at the same site at different periods of time), then statistical inference becomes important.

Like all case studies, these evaluations require a level of context, such as a detailed description of managed care risk arrangements and procedures, that no clinical data source can supply (Hoge, Jacobs, Thakur, and Griffith 1999; Rosenbaum 1997). Collecting this information requires reviewing managed care contracts and procedure manuals and conducting confirmatory discussions with line staff.

LEGAL AND POLITICAL CHALLENGES

Though evaluations based on administrative data have scientific merit, often they can threaten privacy and encounter bureaucratic and/or political obstacles. The number of electronic records that can be accessed and the ease in combining them with other data sources warrants greater concern for research subjects' privacy than traditional medical chart review (Gostin, Lazzarini, Neuslund, and Osterholm). However, since administrative information is collected at no additional burden to the subject, and the risk of unintentional information release is typically very small, federal regulations allow Institutional Review Boards (IRBs) to review administrative data studies on an expedited basis.

Federal regulations for Medicaid data, and state regulations on other data systems (e.g., substance abuse) can place greater limits on data access and handling than ethical requirements of IRBs. These regulations make research based on linked administrative data from multiple agencies difficult. Under such conditions, modifications of the linkage strategies using anonymous data (Haas, Brandenburg, Udvarhelyi, and Epstein 1994; Muse, Mikl, and Smith 1995; Pandiani, Banks, and Gauvin 1997; Pandiani, Banks, and Schacht 1998a) could be useful.

The political difficulties in obtaining data require convergence of evaluation goals with the goals of the host agency, both to justify ethical and legal access to data, and to also assure cooperation from the information staff of the various agencies involved. This convergence maybe difficult to achieve. Public-academic collaborations are necessary, and potentially very valuable to external researchers, and to the behavioral health service agencies themselves.

IMPLICATIONS FOR BEHAVIORAL HEALTH SERVICES

Some of the issues discussed, including lack of client satisfaction measures, level of functioning, and diagnostic details, limit the depth of electronic records as a source for evaluation data. Political and regulatory constraints make these studies even more difficult to implement.

However, most of the problems in conducting these types of studies are solvable. Existing electronic archives are the fastest, cheapest, and often most comprehensive pre-existing source of information on behavioral health service systems. As such, they are invaluable resources for assessing the implementation of managed care polices. Despite their shortcomings, these data can be used to answer the general question of whether a managed care program is functioning adequately and whether it should be considered for replication. Combined with careful qualitative analysis and multiple cases, these data can also help explore the effectiveness of specific managed care techniques.

Many studies using administrative data have already been conducted, but if we hope to make services research more timely and relevant to practice, administrative data can and should be more widely used. In order to do so, behavioral health services researchers and administrators should consider further development and innovation in four areas: (1) improving and standardizing administrative data collection infrastructures; (2) fostering public-academic partnerships to combine operational questions with research questions; (3) better integrating generalizable evaluations into plans for managed care interventions; and (4) refining the theoretical understanding of the effects of specific managed care strategies on behavioral health service system performance and outcome. Progress in these directions will yield benefits for the long-sought goal of data-based management of managed care in the public sector.

ACKNOWLEDGMENTS

This work was funded by the NIMH Training Grant MH19117-08 and the HCFA Dissertation Fellowship 30-P-90662/1-01. The authors thank Howard H. Goldman, M.D., Ph.D., for his comments.

REFERENCES

Adams, M. M., H. G. Wilson, D. L. Casto, C. J. Berg, J. McDermont, J. A. Gaudino, and B. J., McCarthy. 1997. "Constructing Reproductive Histories by Linking Vital Records." *American Journal of Epidemiology* 145(4): 339-348.

Alsop, J. C., and J. D. Langley. 1998. "Determining First Admissions in a Hospital Discharge File via Record Linkage." *Methods in Information Medicine* 37: 32-37.

Bell, R. M., J. Keesey, and T. Richards. 1994. "The Urge to Merge: Linking Vital Statistics Records and Medicaid Claims." *Medical Care* 10: 1004-1018.

Breakey, W. R. 1996. *Integrated Mental Health Services: Modern Community Psychiatry.* New York: Oxford University Press.

Brenner, H. 1994. "Application of Capture-Recapture Methods for Disease Monitoring: Potential Effects of Imperfect Record Linkage." *Methods of Information in Medicine* 33: 502-506.

Brenner, H., and O. Gefeller. 1993. "Use of Positive Predictive Value of Correct for Disease Misclassification in Epidemiologic Studies." *American Journal of Epidemiology* 138 (11): 1007-1015.

Callahan, J. J., D. S. Shepard, R. H. Beinecke, M. J. Larson, and D. Cavanaugh. 1994. "Mental Health/ Substance Abuse and Managed Care: The Massachussetts Medicaid Experience." Unpublished manuscript.

Chang, C., L. Kiser, J. Bailey, M. Martins, W. Gibson, K. Schaberg, D. Mirvis, and W. Applegate. 1998. "Tennessee's Failed Managed Care Program for Mental Health and Substance Abuse Services." *JAMA* 279 (11): 864-869.

Cole, R. E., S. K. Reed, H. M. Babigian, S. W. Brown, and J. Fray. 1994. "A Mental Health Capitation Program. I: Patient Outcomes." *Hospital and Community Psychiatry* 45 (11): 1090-1096.

Dickey, B., S.-L. T. Normand, E. C. Norton, H. Azeni, W. Fisher, and F. Altaffer. 1996. "Managing the Care of Schizophrenia." *Archives of General Psychiatry* 53: 945-952.

Draine, J., and P. Solomon. 1999. "Describing and Evaluating Jail Diversion Services for Persons with Serious Mental Illness." *Psychiatric Services* 50 (1): 56-61.

Galecki, A. T. 1994. "General Class of Covariance Structures for Two or More Repeated Factors in Longitudinal Data Analysis." *Communications in Statistics—Theory and Method* 23 (11): 3105-3119.

Gold, M. 1998. "Beyond Coverage and Supply: Measuring Access to Healthcare in Today's Market." *Health Services Research* 33: 625-652.

Gostin, L. O., Z. Lazzarini, and K. M. Flaherty. 1997. *Legislative Survey of State Confidentiality Laws, with Specific Emphasis on HIV and Immunization.* Atlanta, GA: CDC.

Gostin, L. O., Z. Lazzarini, V. S. Neuslund, and M. T. Osterholm. 1996. "The Public Health Information Infrastructure." *Journal of the American Medical Association* 275 (24): 1921-1927.

Haas, J. S., J. A. Brandenburg, I. S. Udvarhelyi, and A. M. Epstein. 1994. "Creating a Comprehensive Database to Evaluate Health Coverage for Pregnant Women: The Completeness and Validity of a Computerized Linkage Algorithm." *Medical Care* 32 (10): 1053-1057.

Hoge, M. A., S. Jacobs, and N. M. Thakur, 1997. *Understanding Public Sector Managed Care.* Paper presented at the Institute on Psychiatric Services, Washington, DC.

Hoge, M. A., S. Jacobs, N. M. Thakur, and E. E. H. Griffith. 1999. "Ten Dimensions of Public-Sector Managed Care." *Psychiatric Services* 50: 51-55.

Jackson, C. A., G. Sullivan, P. Kogel, S. C. Morton, and M. Fujisaki. 1998. "The Criminal Justice System as a Safety Net." Paper presented at the American Public Health Association Annual Meeting. Washington, DC.

Jennrich, R. I., and M. D. Schuster. 1986. "Unbalanced Repeated-Measures Models with Structured Covariance Matrices." *Biometrics* 42: 805-820.

Leff, H. S., V. Mulkern, M. Leiberman, and B. Raab. 1994. "The Effects of Capitation on Service Access, Adequacy, and Appropriateness." *Administration and Policy in Mental Health* 21 (3): 141-160.

Lehman, A. 1987. "Capitation Payment and Mental Health Care: A Review of the Opportunities and the Risks." *Hospital and Community Psychiatry* 38 (1): 31-38.

Lehman, A. F., and D. M. Steinwachs. 1998. "Translating Research into Practice: The Schizophrenia Patient Outcomes Research Team (PORT) Treatment Recommendations." *Schizophrenia Bulletin* 24 (1): 1-10.

Liang, K.-Y., and S. L. Zeger. 1986. "Longitudinal Data Analysis Using Generalized Linear Models. *Biometrika* 73(1): 13-22.

Lipsitz, S. R., and K. Kim. 1994. "Analysis of Repeated Categorical Data Using Generalized Estimating Equations." *Statistics in Medicine,* 13: 1149-1163.

Lurie, N., I. S. Moscovice, M. Finch, J. B. Christianson, and M. K. Popkin. 1992. "Does Capitation Affect the Health of the Chronically Mentally Ill?" *Journal of the American Medical Association* 267 (24): 3300-3304.

McFarland, B. H., R. E. Johnson, and M. C. Hornbrook. 1996. "Enrollment Duration Service Use, and Costs of Care for Severely Mentally Ill Members of a Health Maintenance Organization." *Archives of General Psychiatry* 53: 938-944.

Mechanic, D. 1998. "Emerging Trends in Mental Health Policy and Practice." *Health Affairs* 7 (6): 82-98.

Mechanic, D., M. Schlesinger, and D. McAlpine. 1995. "Management of Mental Health and Substance Abuse Services: State of the Art and Early Results." *The Milbank Quarterly* 73 (1): 19-55.

Miller, R. H. 1998. "Healthcare Organizational Change: Implications for Access to Care and Its Measurement." *Health Services Research* 33 (3): 653-680.

Muse, A. G., J. Mikl, and P. F. Smith. 1995. Evaluating the Quality of Anonymous Record Linkage Using Deterministic Procedures with the New York State AIDS Registry and Hospital Discharge File." *Statistics in Medicine* 14: 499-509.

Pandiani, J., S. Banks, and L. Gauvin. 1997. A Global Measure of Access to Mental Health Services for a Managed Care Environment." *Journal of Mental Health Administration* 24 (3): 268-277.

Pandiani, J. A., S. M. Banks, and L. M. Schacht. 1998a. "Personal Privacy versus Public Accountability." *The Journal of Behavioral Health Services & Research* 25 (4): 456-463.

Pandiani, J. A., S. M. Banks, and L. M. Schacht. 1998b. "Using Incarceration Rates to Measure Mental Health Program Performance." *Journal of Behavioral Health Services & Research* 25 (3): 300-311.

Reed, S. K., K. D. Hennessy, O. S. Mitchell, and H. M. Babigian. 1994. "A Mental Health Capitation Program. II: Cost-Benefit Analysis. *Hospital and Community Psychiatry* 45 (11): 1097-

Rogers, W. H., K. B. Wells, L. S. Meredith, R. Sturm, and M. A. Burnam. 1993. "Outcomes for Adult Outpatients with Depression under Prepaid or Fee-for-Service Financing." *Archives of General Psychiatry* 50 (7): 517-525.

Roos, L. L., and A. Wajda. 1991. "Record Linkage Studies, Part I: Estimating Information and Evaluating Approaches." *Methods of Information in Medicine* 30: 117-123.

Roos, L. L., R. Walld, A. Wajda, R. Bond, and K. Hartford. 1996. "Record Linkage Strategies, Outpatient Procedures and Administrative Data." *Medical Care* 6: 570-582.

Rosenbaum, S. 1997. "A Look Inside Medicaid Managed Care." *Health Affairs* 16 (4): 266-271.

Santiago, J., and M. R. Barren. 1989. "Arizona: Struggles and Resistance in Implementing Capitation." *New Directions in Mental Health Services* 43 (Fall): 87-96.

Schlesinger, M. 1989. "Striking a Balance: Capitation, the Mentally Ill, and Public Policy." *New Directions for Mental Health Services* 43 (Fall): 97-115.

Smith, M. W. 1998, November 16). "Finding Deceased Subjects by Searching the Social Security Death Index." Paper presented at the Annual Meeting of the American Public Health Association, Washington, DC.

Stroup, T. S., and R. A. Dowart. 1995. "Impact of a Managed Mental Health Program on Medicaid Recipients with Severe Mental Illness." *Psychiatric Services* 46 (9): 885-889.

Sturm, R., C. A. Jackson, L. S. Meredith, W. Yip, W. G. Manning, W. H. Rogers, and K. B. Wells. 1995. Mental Health Care Utilization in Prepaid and Fee-for-Service Plans among Depressed Patients in the Medical outcomes Study." *Health Services Research* 30 (2): 329-340.

Sturm, R., E. A. McGlynn, L. S. Meredith, K. B. Wells, W. G. Manning, and W. H. Rogers. 1994. "Switches Between Prepaid and Fee-for-Service Health Systems among Depressed Outpatients: Results from the Medical Outcomes Study." *Medical Care* 32 (9): 917-929.

Thakur, N., R. Hoff, B. Druss, and J. Catalanotto. 1998. "Using Recidivism Rates as a Quality Indicator for Substance Abuse Treatment Programs." *Psychiatric Services* 49 (10): 1347-1350.

Thompson, J. W., B. J. Burns, H. H. Goldman, and J. Smith. 1992. "Initial Level of Care and Clinical Status in a Managed Mental Health Program." *Hospital & Community Psychiatry* 43 (6): 599-603.

Wajda, A., L. L. Roos, M. Laefsky, and J. A. Singleton. 1991. "Record Linkage Strategies. Part II: Portable Software and Deterministic Matching." *Methods of Information in Medicine* 30: 210-214.

Zeger, S. L., K.-Y. Liang, and P. S. Albert. 1988. "Models of Longitudinal Data: A Generalized Estimating Equation Approach." *Biometrics* (44): 1049-1060.